P9-EDU-898

Intelligence

The psychometry of intelligence is one of psychology's great achievements yet it is poorly understood. Paul Kline's latest book provides a modern, readable introduction to the subject. Written to be clear and concise it none the less provides a rigorous account of the psychometric view of intelligence.

Professor Kline explains factor analysis and the construction of intelligence tests and shows how the resulting factors provide a picture of human abilities. He shows the value of such tests in both applied and theoretical psychology and in so doing answers the critics of intelligence testing. It is one of the few modern texts that deals with the factorial view yet includes modern work in the cognitive field.

The book will be of interest to students of psychology and education, to those taking courses in clinical, educational and occupational psychology, *as well as* those in psychological testing itself.

Paul Kline is Professor of Psychometrics at the University of Exeter. He is one of the leading authorities in the field. His recent books include *A Handbook of Test Construction* and *Psychology Exposed.*

Intelligence

The psychometric view

Paul Kline

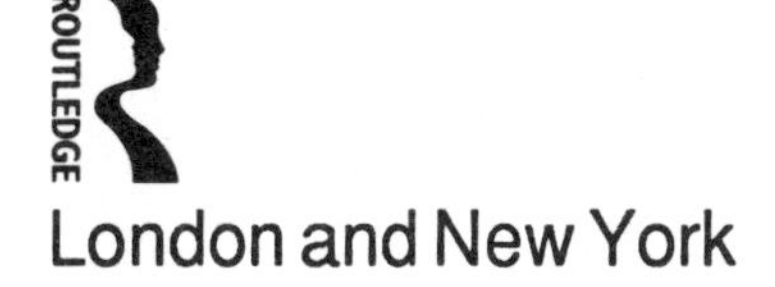

London and New York

First published 1991
by Routledge
11 New Fetter Lane, London EC4P 4EE

Simultaneously published in the USA and Canada
by Routledge
a division of Routledge, Chapman and Hall, Inc.
29 West 35th Street, New York, NY 10001

Typeset by LaserScript Limited, Mitcham, Surrey
Printed in Great Britain

British Library Cataloguing in Publication Data

Kline, Paul
Intelligence: the psychometric view.
1. Man. Intelligence. Measurement
I. Title
153.93

Library of Congress Cataloging in Publication Data

Kline, Paul.
Intelligence: the psychometric view/Paul Kline.
p. cm.
Includes bibliographical references.
1. Intellect. 2. Intelligence tests. 3. Psychometrics.
4. Factor analysis. I. Title.
BF431.K5127 1990
155.9 – dc20 90-32231

ISBN 0-415-05511-3
0-415-05512-1 (pbk)

Contents

Figures and tables

Figures

Tables

Preface

The purpose of this book is to restate the psychometric view of intelligence which, for a variety of reasons, intellectual and social, has fallen into obscurity. Intellectually the rise of cognitive psychology and cognitive science has caused psychologists to stress processes of thinking and to regard intelligence as a somewhat useless, global label. Among the social factors which have led in some cases to direct attacks on the concept of intelligence are the generally egalitarian zeitgeist, claims that intelligence testing leads to racial discrimination or even reflects it, and the unmasking of one of the leading practitioners of intelligence testing, Sir Cyril Burt, as a fraud.

In addition to this the notion of intelligence as even a partly heritable characteristic is ill at ease in Thatcherite society where failure to succeed is regarded as a moral flaw, hard work and determination being the necessary ingredients. This accords well with the views at the other end of the political spectrum where failure is seen as resulting from the oppression and enslavement of the poor by the rich.

In this book I try to put the record straight and present the latest psychometric findings concerning the nature of intelligence and other human abilities. I am encouraged in this endeavour by the fact that in the real world, beyond the narrow boundaries of academic psychology, intelligence tests and other psychometric methods of assessment are being used in ever greater numbers as employers seek to find people best suited to their jobs. The application of psychometrics is one of the few technological successes in psychology and it is right that students and practitioners should understand the theoretical bases of this work. In addition to this a knowledge of the psychometrics of ability is valuable for a wider understanding of the psychology of human development and achievement.

Chapter one

The meaning of intelligence

In this chapter I shall discuss the meaning of intelligence as it is generally used in English. This definition will be compared with that given it by psychologists. Finally I shall examine the status of intelligence as a concept since it has been argued that it is simply redundant and hence deleterious for the scientific study of human abilities.

As Jensen (1980) has pointed out, the notion of intelligence can be found in the great texts of the Hindus and the Ancient Greeks. This is hardly surprising since in almost every activity we can see things being done intelligently and otherwise. Intelligence is popularly defined as the ability to learn, understand and deal with novel situations. The intelligent person is seen as quick-witted, acute, keen, sharp, canny, astute, bright and brilliant, to select just a few of the relevant adjectives in everyday language. At the other end of the scale the unintelligent person is described as dim, dull, thick, half-witted or stupid. These adjectives give the flavour of intelligence as the word is used in English.

An example, perhaps, will bring the concept to life. The philosopher Bertrand Russell illustrated most of these characteristics to a very high degree. When he spoke on a subject, however complex, he was wonderfully lucid, picking out the essentials of the problem. His arguments were clear and convincing. His replies to objections were almost always devastating. He could instantly see the flaws and errors in reasoning and arguments. His superb intelligence, indeed, was the basis of his *Principia Mathematica* (Russell, 1911).

Now this example highlights some other important points about intelligence. Russell was distinguished for the analytic quality of his reasoning and his quickness at seeing the point. This is quite different from massive learning and prodigious memory, although these may be found together. In universities there reside many scholars who are hugely knowledgeable, in their fields, but are not highly intelligent in the sense discussed above. In my view it is such individuals who bring learning and scholarship into disrepute. The unintelligent scholar simply flounders in a sea of facts and contributes little to the world. The scholars we remember combine intelligence and learning.

I think it is clear, from these introductory paragraphs, how the term intelligence is conceptualised in English. Psychological definitions, although they attempt precision, are highly similar although some are remarkably uninformative. In an effort to be comprehensive, for example, Brown (1976) writes 'intelligence, to the psychologist, is the ability to perform certain types of task'. Without specifying those tasks this is not satisfactory. This definition is similar to that given by Boring (1923) who claimed that intelligence is what intelligence tests measure. This is not so circular as it first appears since, as will be discussed in later chapters of this book, psychometrists are able to specify empirically what intelligence tests do measure. Some psychologists have attempted more precise definitions, for Boring simply changed the problem into having to define what intelligence tests measure. Terman, the author of one of the best known intelligence tests, the Stanford-Binet test, defined it as the power to think abstractly, to be self-critical and to be adaptive. Wechsler (1975), the constructor of the WAIS and WISC tests, defined intelligence as the capacity of a person to understand the world and meet its demands, a definition which is remarkably similar to the common English usage. Humphreys (1975) has defined intelligence as our repertoire of intellectual skills. However, this again throws the definition back – to the meaning of intellectual skills.

Several points about this sample of definitions by psychologists deserve comment. First there is no clear agreement. Thus a dictionary of psychology could offer no simple definition. Despite this there is some concurrence, namely that intelligence is a capacity of some kind that is useful in problem solving. In this respect, as I have argued, it is little different from the common-sense definition. Such a general and rather diffuse definition is hardly suitable for scientific study and I now want to turn to definitions that are not just verbal but are more experimental and empirical. These are part of the branch of psychology known as psychometrics, measuring the soul. Would it could be done. This is an aspect of psychology which originated in Great Britain and in which intelligence was a fundamental concept. I should point out, at this point, that in many branches of science purely verbal definitions are not useful. For example it would be possible to decide arbitrarily to define intelligence as abstract reasoning ability. However, if it were the case that, in the real world, intelligence, as it was used, was more than this component, such a definition, no matter how consensual, would be useless. This is simply definition by fiat which in empirical science will never do as exemplified in astronomy by claims that the earth was flat and that the sun went round the earth.

In this book I intend to explicate this psychometric concept of intelligence which often now lies neglected, despite the formidable research which underpins its claims. As a beginning I shall discuss the psychometric definition but before this can be done I shall have to make some general points about psychometrics and its methods.

Essentially psychometrics was the invention of Spearman at the turn of the century in University College London. His paper ' "General intelligence": objectively determined and measured', published in the *American Journal of Psychology* in 1904 contains all the basic psychometric reasoning about the nature of intelligence. In discussing this pioneering work the psychometric definition and concept of intelligence will become clear. The question Spearman asked was this: why is it that human abilities are positively correlated, that is why is there a general tendency for those who are good at one thing to be good at others? This is the phenomenon that teachers refer to as correlation not compensation. It is, alas, not the case that the dimwit at maths is good at woodwork. The simple answer, and as it has turned out too simple, as will be fully discussed in later chapters of this book, was that this correlation could be accounted for by general intelligence or *g*. This *g* was common to all tasks requiring ability, hence the correlation. For example ability at French depended upon *g* plus a specific French component; physics depended upon *g* plus a physics component and so on. Spearman's achievement was to invent a statistical method, factor analysis, which could uncover this structure of abilities.

Later psychometrists, right up to the present day, have, in fact, developed this approach. Factor analysis has been extensively improved and made more efficient and this technique will be fully described in Chapter 2. Far more abilities have been sampled and the empirical nature of the *g* factor or factors has been explored in the clinical, educational and industrial fields. In addition there has been considerable theorising about its nature and all these results can be found in the relevant chapters of this book. Suffice it to say now that in psychometry, intelligence is equated with this general ability or *g* the factor common to all problem-solving abilities. Since the best intelligence tests are deliberately constructed to measure this *g* factor it does make sense to define intelligence as what intelligence tests measure, provided that the *g* factor can be specified.

The nature of the *g* factor

Modern factor analysis (Cattell, 1971) has broken the *g* factor into two: crystallised and fluid intelligence. Fluid intelligence is the basic reasoning ability of an individual dependent, to a large extent, on the neurology of her brain. Crystallised intelligence represents this ability as it is evinced in the skills valued by the culture in which the individual lives. In the West, therefore, fluid ability is invested in science and technology and not at all in hunting and tracking, to give a cross-cultural example. Factor analysis, as will be discussed in later chapters of this book, has revealed the extent to which these *g* factors are implicated in different tasks and occupations. Skill in all activities which demand problem solving is correlated with *g*, highly in some cases but less so in others. Nevertheless the relationship is always positive. This is the broad, general picture of intelligence as conceptualised in psychometrics. However, there have been many objections to this psychometric concept and the most important of these will be discussed briefly in this introductory chapter.

A recent objection to the factor analytic work in human intelligence has been raised by cognitive psychologists, for example Hunt (1978), Carroll (1983) and Sternberg (1977). This concerns the utility of the *g* factor or factors. Even if it is admitted that a common factor can be found in many problem-solving tasks, labelling it as *g*, or reasoning ability, or any other descriptor, does not explain it. The cognitive processes involved in problem solution are not thereby understood. At best the factor is descriptive. What is necessary, therefore, is a study of the processes employed in problem solution. This is an important point and in Chapter 9 modern work, where the factor is explicated experimentally, will be discussed. However, it must be pointed out here that this research is an extension of the factor analytic work, not a replacement, and could not take place at all if ability factors had not firmly been established. Certainly the main researchers in factor analysis, such as Cattell, Guilford and Eysenck, regard such studies as essential parts of their research programmes (Eysenck, 1982).

The second important objection concerns the status of intelligence as a concept. Miles (1957) has argued in the behaviourist tradition that intelligence is a redundant concept. We say, the argument runs, that someone is behaving intelligently because he has a lot of intelligence. This is, however, a circular argument because the only evidence that intelligence exists comes from the observation of intelligent behaviour. This argument has been widely applied to the notion of all psychological entities by Skinner (1953). However, it will not do. Intelligence can be measured independently by tests and the scores can be used to predict a disparate variety of outcomes in both the educational and occupational spheres, as will be fully discussed in later chapters of this book.

An example will clarify the point. If I required a number of people to learn Classical Greek to a high standard within a few months, I would select those who performed well on a test of *g* such as Raven's Matrices (Raven, 1965c), a test which is heavily *g* loaded and which consists of sequences of patterns which have to be completed. This test has nothing in common, it will be noted, with Greek, in respect of content. I use intelligence as an explanation of their superior performance at Greek, compared with moderate scorers on the test. This use of an explanatory concept, such as intelligence, which can be measured, is no different from the use of concepts in other sciences. An analogy can be found in the case of deafness. It is perfectly sensible to explain an individual's failure to learn well in school by invoking deafness as a cause. The fact is that there is no error in the proposition of abstract constructs if they can be measured independently of the observations which they are meant to subsume. This, however, is not to deny that it could be valuable to investigate cognitive processes underlying intelligence, as has been discussed. Finally, it must be pointed out that the status of intelligence as an explanatory concept in this example does not stand or fall by the fact of a low or high correlation with Greek. If the correlation is low then the explanation is wrong. My point here is simply that the status of intelligence as an explanatory concept is perfectly respectable.

Howe (1988) has attempted to demonstrate, by surveying a variety of claims about intelligence, that it is an empty concept. Some of these are relatively trivial and I shall say nothing about them. Some of the arguments are important and I shall scrutinise these briefly. One of the difficulties in analysing these arguments lies in the vagueness of the language. For example he asks whether level of intelligence is controlled by inherited mechanisms? But what would constitute a positive answer? Genes are either implicated or not. Later he writes 'the view that knowledge that a hereditary cause of variability in human intelligence may exist must provide any strong clue to the precise direction of form of that cause is quite wrong.' This is indeed unanswerable because the meaning of 'precise direction of form of that cause' is unclear. Indeed it seems empty of sense. The next sentence is no better: 'it has been supposed that any effects of genetic factors would need to be exerted via the kinds of physiological processes that determine the power, speed or flexibility of physiological processes underlying cognition . . .' This he claims is wrong. However, why should there be processes underlying processes in the first place? Why could the action not be direct? This argument seems to be loose in the extreme. Howe continues 'for instance hereditary factors could exert their mark on intelligence by affecting motivational processes of some kind or personality variables . . . or sensory perception.' These arguments, while possible, are empirically false as the extensive work of Cattell in the field of motivation, personality and ability demonstrates. Personality and motivational factors are independent of the ability sphere (Cattell, 1981). There can be no question here that Howe has failed to make his point. In a later chapter the work on the hereditary determination of the *g* factors will be examined.

As a final riposte Howe argues that the extraction of a factor from a set of intellectual tests tells us little if anything about the causes of intellectual variability. No defence of this extravagant claim is offered and in the following chapter where the rationale and methods of factor analysis are set out, it will become obvious that this view is mistaken. Similarly the arguments used by Alice Heim (1975), who was a brilliant constructor of intelligence tests, in connection with the unreliability of factor analytic solutions, will be dealt with in this chapter. Suffice it to say here, that modern factor analyses can be made highly reliable and replicable.

From the introductory discussion of this first chapter, I hope it is clear that the psychometric concept of intelligence which I shall explicate in detail in the remaining chapters of this book is similar to that in common English usage. It can be roughly seen as a general reasoning capacity useful in problem-solving tasks of all kinds. Objections concerning its redundancy and its lack of explanatory power have been shown to be ill conceived and without force. Other objections which have been raised about psychometric *g* will be discussed in the relevant chapters of this book.

Chapter two

Factor analysis

In this chapter I shall attempt to set out the rationale and method of factor analysis so that it can be properly understood. The emphasis will not be upon the algebra or computation (which in any case are now the province of the computer and which in their refined form give little indication of their logic and meaning other than to fine mathematicians). Rather I shall concentrate upon the logic and function of factor analysis and upon the meaning of the technical terms which have to be used to describe the results. Factor analysis is open to many abuses of logic and sense all of which can be found in published reports now that any social scientist, however innumerate, can use the method. All these errors of method and interpretation will be discussed. Without this the psychometric concept of intelligence and human abilities in general cannot be grasped and there can be no doubt that many of the objections which have been raised against psychometric work are ill founded on account of just such misunderstandings.

Before I begin this explication of factor analysis, for those readers who require a clear account of all the computational procedures of the different methods, Cattell (1978) and Harman (1976) can be recommended. Child (1970) is useful because his is the most simple account of some of the computational methods.

Definition of terms in factor analysis

First I shall define the terms of factor analysis because, without this, discussion is hopelessly rambling. I shall begin with some basic statistical definitions.

Variable

Any characteristic on which individuals or the same individual over time can vary. Height, weight and test scores are obvious examples.

Variance

This is the variation in scores on a variable of a sample or population. Sample and population must be distinguished. If we are studying reading among 10-year-old children in Great Britain the population consists of all of these children. In almost all research we are forced to use samples and it is important that these samples are representative of the populations which they purport to represent. As we shall see sampling is important in factor analysis.

Correlation

The correlation coefficient, *r*, indicates the degree of agreement between two sets of scores. If high scores on one variable are associated with high scores on the other the correlation is positive, as is the case with all human abilities. If high scores on one variable are associated with low scores on the other, the correlation is negative. The correlation coefficient runs from 1 to minus 1. A correlation of 1 indicates perfect agreement, minus 1 the opposite, for example if the order of one variable was a perfect inversion of the other. A correlation of 0 shows that agreement between the variables is random, that is there is no relationship at all.

Meaning and interpretation of correlations

If we square a correlation coefficient it indicates the amount of variance in common between the two sets of scores. Thus a correlation of 0.8 shows that there is 64 per cent of variance in common on the two variables. However, this figure still requires some interpretation if it is to be given any psychological meaning. There are various determinants of correlations and these will now be discussed.

Common elements may be one cause of a substantial correlation. For example there is a high correlation between scores on Latin and Greek. This is because there are similar abilities involved in both tasks. High *g* is important to understand the grammatical rules; verbal ability is required for both, especially in elegant translations. In addition there is a common element of interest in the academic questions of the ancient world, in contrast, say, to the vocational aspirations of trainee accountants.

A cause common to both variables may account for the correlation. For example there is a correlation between cigarette smoking and lung cancer. Eysenck (1980), however, has maintained that this is accounted for by a third factor which causes an individual to smoke and to be liable to lung cancer. This is a possible interpretation of the correlation and it highlights an important point that is often ignored. Causation cannot be inferred from correlation. Since the Second World War there is a correlation between imports of luxury goods and increases in the mean height of adults in Great Britain. This, however, results from increases in prosperity for the mass of the population and is clearly not causal.

Magnitude of correlations

The magnitude of a correlation reflects the degree of variance in common between two sets of scores. However, this can be affected by what are essentially distorting variables. Two factors are particularly influential and must always be borne in mind when scrutinising correlations. Since correlations are the basis of factor analysis this may also be similarly distorted. Correlations can be reduced in size by homogeneity of variance. If we take the correlation between intelligence and academic success across the whole range of ability it is likely to be substantial, around 0.5. However, this would include those of such low intelligence that they cannot read or write and would, necessarily, score 0 on any academic test. However, if our sample is selected for intelligence (as for example at a good university where all students have IQs beyond 120) then the correlation is bound to fall. Everyone has sufficient ability to do the work and thus other factors, such as special abilities, interests and perseverance, become important. This point becomes absolutely clear if we were to imagine the impossible case where every student had the same IQ. There its relationship with any other variable would be 0.

The unreliability of measures is the other source of reduction of correlations. The reliability of a test is essentially its correlation with itself. Ideally this should be unity but in practice it falls far short of this. The reliability is the upper limit of the possible correlations with a test. Thus if both measures are unreliable correlations are necessarily reduced. Fortunately most good intelligence tests have high reliabilities but in other fields, such as personality, this is not so, and great care has to be taken in interpreting any results.

Factor analysis

Factor analysis is a statistical method in which variations in scores on a number of variables are expressed in a smaller number of dimensions or constructs. These are the factors. In almost all psychological, psychometric studies factor analysis is applied to the correlations between variables. The factor analysis shows the correlations of each of the variables with these constructs or dimensions. Factors are defined by these correlations, called factor loadings. An artificial example will clarify the point.

Suppose that intercorrelations had been obtained between all the subjects commonly studied at school. A three-factor solution would undoubtedly account for most of the intercorrelations. One factor would be general with loadings on all the subjects. There would be high loadings on subjects such as Latin, maths and physics and much smaller ones on physical education, woodwork and domestic studies. This would be identified from this pattern of correlations as *g*, or general ability. Complex, 'hard' subjects have the highest loadings on this intelligence factor. A second factor would load on English, French, German, history, geography (but less) and on all subjects where language skills were important. This is the verbal ability factor. A third factor would emerge, which loaded on maths, physics, chemistry and statistics, and would be a factor of numerical reasoning. The identification of these factors springs from their loadings. What is a construct that correlates with scores on language tests? Verbal ability is a sound inference. In psychometric work these factors would be further studied experimentally in new research in order to validate the identification based upon their loadings.

This example of a factor analysis illustrates the powerful ability of the technique to simplify complex data. If we were to examine the correlations between all the variables there would be far too much information for the mind to take in. However, three factors can be grasped. The scores on these three factors would be virtually as good an indication of the ability of the individuals who had taken the tests as the original scores on all the variables.

Definition of a factor

The definition of a factor adopted in this book is that of Royce (1963) and is one that is used, in practice, by the majority of psychometrists. In fact it is the definition which I used above in my discussion of factor analysis. A factor is a construct operationally defined by its factor loadings.

Factor loadings

These are the correlations of the variables with the factor. In my school example the verbal factor was identified by the fact that it had loadings on tests that clearly required verbal abilities, and, equally importantly, had no loadings on tests that did not require such ability.

It is useful to consider these two definitions together, since it is hardly possible to talk of factors apart from their loadings. The first point to notice concerning this definition of factors is that their status as constructs is very different from that of the same terms in everyday language. To take up the school example again, many people, entirely ignorant of factor analysis, might suppose that verbal ability was important in the learning of a language. However, without factor analysis there is no evidence, except from common sense, that such a construct is anything more than an abstraction based upon the fact that some people learn languages more quickly than others. Furthermore there is bound to be disagreement concerning the definition of so subjective a construct. The factor analytic construct is defined by its loadings with numerical precision. Furthermore the existence of the factor guarantees that it is a useful construct that can account for the variance in the test scores.

Identification of factors

Although a factor is defined by its loadings such a definition is not sufficient in itself. This is because in some cases the mix of variables on the factor is such that one cannot be certain from the loadings alone how the factor should be identified. A more important reason for requiring further evidence is the argument, used by Heim (1975) and others who do not favour factor analysis, that factors are simply mathematical abstractions derived from correlations and that they bear no necessary relationship to anything in the real world, beyond the correlation matrix.

If we think, for example, that we have a *g* factor then we would carry out further studies to demonstrate the case. We would compare the factor scores of brilliant scholars and others whom we considered to be highly intelligent with controls. We might give the *g* test to those beginning a course and show correlations with success, and so on. External evidence is necessary to identify a factor convincingly.

Communality

An important question concerns the adequacy and quality of a factor analysis. As we have seen, a factor analysis represents the variance among the variables with a smaller number of factors. One test is to examine the communality of the variables, defined as the proportion of variance accounted for by the factors. An example will clarify the issue. Suppose that we have a three-factor solution. To calculate for each variable the variance accounted for by the factors, the common variance, we square and add the loadings on each of the three factors. So that if Raven's Matrices loaded 0.7 on factor 1, 0.2 on factor 2 and 0.0 on factor 3, the communality would be 0.53. Thus the factors had accounted for 53 per cent of the variance in the test. This, therefore, is not a good solution for understanding the variance in this test. The other part of the variance in the Matrices (47 per cent), unique variance, consists of specific variance plus error variance. Specific variance is of little psychological interest being restricted to a test. It should be pointed out that these figures will change in different studies. With different variables more factors might be extracted and thus the common variance would be larger. From this it can be seen that one test of the adequacy of a factor analysis is that the communalities of the variables are large. Ideally they should all be beyond 80 per cent.

Reproducing the correlations

Another test of a factor solution is to see how well the original correlations between the variables can be reproduced from the factor loadings. Simple cross-multiplication of the loadings yields the correlations. If these are way off the original observations the factor solution is poor. The more accurately these correlations can be reproduced, the better the solution. Cattell (1978) has an easy worked example. Of course, the fact that the correlations can be reproduced from a smaller number of factors indicates their power in understanding the variance within the correlation matrix.

Eigen values or latent roots

These indicate the size of each factor, the amount of variance in the matrix for which it accounts. They are calculated by summing the squares of each loading on a factor.

Factor rotation

Factors can be conceptualised as axes in factor space, as shown in Figure 2.1. It is clear that, by rotating the factors, factor loadings change although all these positions are mathematically equivalent to each other. This fact that there is an infinity of mathematically equivalent solutions, depending upon where the rotation of axes stops, gives rise to a number of severe problems.

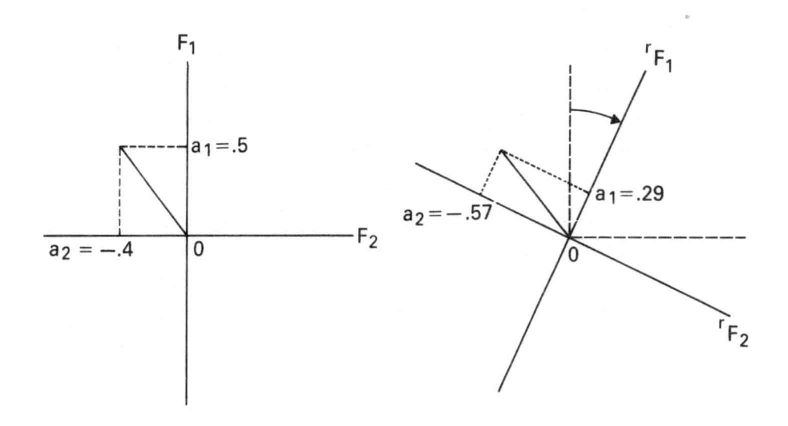

Figure 2.1 Orthogonal rotation and changing factor loadings
Source: Adapted from Cattell (1978)

Problems in factor analysis

If factor analysis, with its capability to extract the underlying most important dimensions from a set of measurements, is so powerful a technique as would appear from the initial discussion and description, then it is fair to ask why it is not more widely used. In fact it is virtually restricted to psychometrics and test construction and many psychologists know nothing of it. There are various reasons for this which must now be examined.

The major difficulty stems from the fact that there is no single unique solution to a factor analysis. There is an infinite number of numerically equivalent solutions, as was discussed in our definition of rotation. Since there is no a priori reason to select one solution rather than another, most experimentalists have concluded that the method is not valuable.

That this is no theoretical problem is attested by the fact that in the psychometrics of both ability and personality there appears to be little agreement as to the best or the most useful solution. In ability the original workers, Spearman and Thurstone, had very different solutions. The former had one general factor and a number of specifics, the latter nine primary factors. Even modern workers disagree considerably. Guilford (1967) has claimed 120 ability factors, while Cattell (1971) prefers to deal with 5 large factors. In the field of personality there are similar disagreements between factorists. Heim (1975) in the experimental Cambridge tradition has declared a plague on all your houses and condemned the method.

Yet modern factor analysts have overcome all these problems and replicable and meaningful solutions can be obtained. How this can be achieved and the reasons for the disagreements among different factor analyses will now be discussed. Essentially these difficulties are logical and technical. I shall begin with the logical problems.

Logical problems

One cause of disagreement can be quickly dealt with. This concerns the identification of factors where there is general agreement that the factors do emerge. The best example of this is the neuroticism factor in the field of personality. Cattell (e.g. Cattell and Kline, 1977) and Eysenck (Eysenck and Eysenck, 1975) both agree that there is a large factor which loads on a range of variables related to neurotic disorders and psychiatric disturbance. Furthermore empirical research studies indicate these are certainly identical factors (Kline and Barrett, 1983). However, for theoretical reasons, Cattell labels the factor Anxiety, while Eysenck calls it Neuroticism. In fact the nomenclature is unimportant both from the viewpoint of understanding the structure of personality and measuring the variable for application in educational, clinical and industrial psychology.

Importance of simple structure

There is, however, a much more serious problem arising from the fact that there is an infinity of possible solutions. How is one solution to be chosen? This problem was answered, logically, at least, by Thurstone (1947). Each solution can be regarded as a set of hypotheses to account for the correlations between the variables. In the scientific method it has generally been agreed that, among competing hypotheses, the simplest is the best, the law of parsimony or Occam's razor. Thurstone, therefore, aimed to rotate solutions to simple structure, simple structure being by definition the most simple solution. The essence of simple structure thus defined is to arrive at a solution which maximises the number of zero or near zero loadings. In this way each factor is necessarily simple since it has a few high loadings. The elegant logic of Thurstone's answer to the problem has been supported empirically by Cattell (1978) who demonstrated that simple solutions are replicable and yield meaningful factors in cases where the factors are known. Both these points will be discussed further in later sections of this chapter.

Thus the logic of factor analysis is clear. The simple structure rotation, as the most parsimonious account, is the solution to be chosen. Unfortunately, however, as Cattell (1973) has demonstrated, although there is reasonable agreement among the leading exponents of factor analysis that simple structure is the aim, there is little agreement how this should be obtained. Furthermore and this is, perhaps, even more serious, Cattell (1973, 1978) has shown that simple structure can never be obtained unless the factor analysis is technically adequate and that the majority of factor analyses are methodologically flawed. It is these technical errors that have led to the disparity in findings and thus it is necessary to set out briefly the technical necessities of good factor analysis and the concomitants of the many different types of error.

Technical problems of factor analysis

Sampling of variables

If we are trying to map a complex field, as is often the case with factor analysis, a technique known as exploratory analysis, then it is essential to sample the whole population of variables. In the field of ability, for example, if no measures of an ability are included in a factorial analysis, no factor could possibly emerge and our picture of abilities would be bound to be wrong. In fact many human abilities, especially those requiring specialised skills, have no tests and have not been subjected to factor analysis. One example would be the skill of testing cheese for ripeness, a skill which has proved difficult to automate.

In connection with the sampling of variables it should be noted that at least three variables are required to mark a factor. Thus if we suspect that there is a factor of musical ability a minimum of three tests is required. From this it should be obvious that where the full field of variables has not been sampled the resulting factor analysis is bound to be less than complete.

Sampling of subjects

In exploratory analyses, especially, it is essential that the sampling of subjects is such that all variables have their full variance. If variance is restricted the factor cannot emerge with any clarity. A common example of this is in the study of university students where there is a restricted range of intelligence thus affecting the *g* factor.

The numbers of observations and variables

It is essential for mathematical reasons in the matrix algebra of factor analysis that the number of subjects must exceed the number of variables. If it does not the results are affected by mathematical artefacts, as has been well documented by Nunnally (1978). However, concerning the size of the subject to variables ratio there is considerable dispute. Nunnally is the most conservative authority, claiming that there should be ten times the number of subjects to variables. At the other extreme is Guilford (1958) who has argued for a ratio of 2:1.

What is noteworthy about these claims is that there is no rationale to support them: they were devised from the experience of their authors. Barrett and Kline (1981) investigated this question using items from two personality tests, Eysenck's EPQ and Cattell's 16PF test. With a ratio of subjects to variables of 20:1 so clear a factor structure emerged that this was used as a baseline to evaluate results from smaller samples. It was found that the main factors emerged with the ratio as low as 2:1 and that at 3:1 the differences from the large sample were negligible. The influence of the actual number of subjects was also studied. Here it was found that a large *N* improves the reliability of the results almost certainly because the standard error of the correlations is thereby reduced. Factor analyses with samples of fewer than 100 subjects must be treated with caution and all such factorings require replication. A ratio of 2:1 would appear satisfactory although it must be stressed that this study used data of unusual clarity. Certainly an observation to variable ratio of 3:1 is sufficient.

Principal components or principal factors

Modern computing packages usually offer this choice (at least) of initial factoring method. The most important difference between these two procedures, if we ignore the fact that strictly principal components analysis is not factor analysis because it produces as many components as variables, is their treatment of the variance in the correlation matrix. Principal components analysis uses all the variance, including error variance; principal factors or principal axes analysis, however, excludes the error variance by estimating communalities. Carroll (1983) has argued that this makes principal factors a more accurate method but in large matrices of more than 25 variables Harman (1976) has shown that there is almost no difference in the factors. Given this, the distinction is of little significance but with smaller matrices principal factors analysis is to be preferred.

Maximum likelihood factor analysis

This method of factor analysis which requires extensive computing facilities differs from principal components and principal factors in that it has a statistical rationale which allows one to test for the significance of the number of factors. It is, therefore, a method which is greatly favoured by statisticians as distinct from psychologists. However, even these statistical advantages are not as great as they might first appear because the statistical test for the number of factors in large matrices is not always sensitive enough to choose between different possibilities.

Despite the claims made for the advantages of each method of initial factoring, it appears to be the case that with large numbers of factors the differences are trivial. This is particularly so because not all the factors emerging from an analysis are of psychological significance and the choice of how many should be rotated is crucial to obtaining simple structure.

How many factors should be rotated?

Cattell (1978) has argued that this is one of the most critical issues in reaching simple structure, as mentioned above, because if too many factors are rotated, factors split and in consequence lose generality. If too few factors are put into the rotation a few general factors are produced and dimensionality is compressed.

Kline and Barrett (1983) have examined this problem in considerable detail but some relatively straightforward conclusions can be drawn. First they showed in their empirical studies of different rotational procedures (Barrett and Kline, 1982) that there was no one best method for choosing the number of factors. However, there was no doubt that the criterion of rotating factors with eigen values greater than 1, a method used in many of the factor analytic computing packages, is unsatisfactory. As Cattell (1978) argued, this greatly over-estimates the number of factors, especially in large matrices where many of the factors account for relatively small amounts of variance. Two tests seemed to work well and to be in general agreement. These were the Scree test (Cattell, 1966) and the Velicer test (Velicer, 1976). The Scree test is highly subjective and depends upon the experience of those who use it, although over the years, working with colleagues, there has been little disagreement. Barrett and Kline (1982) developed an automated version of the Scree test, although there is some subjectivity even in this. The Velicer test is objective but there is far less evidence as to its validity. I think it best if both these tests are administered and both solutions rotated if these are different. The actual choice should be made on a priori, subjective grounds. Effectively, methods to select the number of factors to be rotated should be regarded as guidelines rather than rules.

Orthogonal or oblique rotations

In our definition of factor rotation it was explained how factor axes could be rotated to any position and that the aim of factor analysts was to produce a simple structure solution. First a few terms should be defined.

Orthogonal rotation

In orthogonal rotation factor axes are at right angles to each other and are uncorrelated.

Oblique rotation

Here the factors are oblique, at less than right angles to each other. They are, therefore, correlated since the cosine of the angle between them gives us the correlation. That this should be so can be seen if we imagine, in Figure 2.1, that we rotated factor 2 to the same position as factor 1, where, of course, it would become identical.

If we are aiming at simple structure, then the question arises as to whether factors should be oblique or orthogonal. There are two arguments to be considered in the light of the definition of simple structure as the most simple solution. Thurstone, it will be remembered, sought factors with a few high loadings, the majority being around zero. With this operational definition it is highly likely that the oblique solution will provide the simplest structure because the axes can take up any position and are not restricted by orthogonality. Furthermore, as has been argued by Cattell (1978) and Kline (1979), there is an additional argument in favour of oblique rotations, namely that, in the real world, it is not unreasonable to think that factors, as important determiners of behaviour, would be correlated. These two arguments seem overwhelming and the only factorist of any note to oppose them is Guilford (1959).

His argument turns on the meaning of simple. He claims that, while it is true that oblique factors may individually be more simple than their orthogonal equivalents, in fact as a hypothesis a set of correlated factors is more complex than a set of orthogonal factors. Hence the preference for orthogonal rotations. While there may be some force in this argument, empirical work with factor analytic solutions supports the notion that oblique solutions are the more simple. In my view the oblique solution is to be preferred, as a means of reaching simple structure.

Higher order factors

There is a further reason for arguing that oblique factors are more useful in providing simple accounts of the variance in correlation matrices compared with orthogonal factors. This stems from the fact that it is possible to factorise the correlations between factors and thus produce higher order factors. This can be done with the primary ability factors of Thurstone, for example, and at the higher order *g* emerges, loading on the primary factors.

From this discussion I think that it must be concluded that oblique rotations are to be preferred because they yield more simple factors and because, especially in the sphere of abilities, where, as we have seen, there is a positive manifold, it is likely on theoretical grounds that factors will be correlated. If the aim is simple structure then oblique rotations are more likely to reach it.

Group and specific factors

We have mentioned group and specific factors in this discussion of factor analysis. However, it is now necessary to examine them in a little more detail. A group factor is defined as one which loads on a number of variables while a specific factor loads only on one variable, i.e. it is specific to it. This appears perfectly clear. Unfortunately, as Cattell (1978) has argued, this distinction is over simple. Thus it is possible if we find a specific factor, in a factor analysis of items, to write a large number of similar items and thus produce a group factor loading on these items. However, this is not a true group factor, according to the argument, but a bloated specific. How, then, are group factors to be discriminated from them? Fortunately there is a simple answer to this question: by correlations with external criteria. A bloated specific is still a specific and thus will not correlate with any criterion. This is not the case with group factors. This distinction between group and bloated specific factors is important when we come to examine the substantive findings of factor analysis.

Methods of rotation

Cattell (1973, 1978; Cattell and Kline, 1977) again and again makes the point that one of the major causes of failure to reach simple structure, and thus to produce replicable and reliable factors, is poor rotational procedures, even when these are designed to produce oblique factors. This is a complex problem especially because a large number of different rotational procedures have been developed. Nevertheless there is now some agreement as to how rotation should be carried out and I shall summarise the conclusions. I must point out here that this is not an academic exercise designed to test the patience and logic of the reader. When we come to discuss the substantive findings of the factor analysis of intelligence we shall be able to see how some research may be discounted, and some claims treated with extreme caution.

Hyperplane count

To understand our discussion I shall have to define one further term. Hyperplanes are the boundaries alongside factors. If we fix these as + or –0.05 (thus regarding all loadings of this size or less as zero) then the hyperplane count is the number or percentage of factor loadings within these hyperplanes. To maximise the hyperplane count, therefore, is to maximise the number of zero loadings in the factor analysis. Now, the definition of simple structure, it will be remembered, was the set of factors with a few high loadings and the rest zero.

Thus we can now see why the best rotational procedures aim to maximise the hyperplane count. This inevitably produces simple structure. Hakstian (1971), in his study of different rotational procedures, found that the most efficient and thus the one best at obtaining simple structure was Direct Oblimin. The more simple Promax rotation is also efficient where the factors are not too oblique. Both these rotations are available on the leading computer packages.

The technically adequate factor analysis can now be described. There will be proper sampling of variables and subjects. The principal factor analysis (or components within a large matrix) will be subjected to an oblique factor analysis, in which significant factors, as selected by the Scree test or Velicer test, are rotated to simple structure by a procedure which maximises the hyperplane count, preferably Direct Oblimin. Such a factor analysis should yield factors which are replicable and which account for much of the variance in the matrix.

The status of simple structure factors

Cattell (1978) following Thurstone (1947) has stressed the importance of obtaining simple structure for reasons which have been discussed, namely that it is the simplest solution and that it is replicable. There are other important points about simple structure factors and these deserve scrutiny.

Cattell (1978) argues that simple structure factors not only are replicable, and this is important given the infinity of possible solutions, but also are causal determiners. This argument springs from two sources. If random data are factored, simple structure, as defined by the hyperplane count, cannot be reached. More important than this, some artificial data sets have been factored, data sets where there are known causal agencies. One example concerned the behaviour of balls. Simple structure analysis showed that this could be predicted from three factors: weight, size and elasticity. Such examples indicate the hopelessness of attempting to regard factors as merely algebraic descriptions. This is the basis for the simple structure factoring of human abilities that important determiners of the variance and covariance can be discovered.

Factor structure and factor pattern

Two small points deserve a brief mention. In oblique solutions, a factor pattern and a factor structure are produced. Examination of these reveals that they are always highly similar, so similar, indeed, that substantive interpretation of the factors would not differ. In the orthogonal case they are identical. The factor structure loadings are the correlations of the variables with the factor, the definition of factor loading earlier in this chapter. The factor loadings in the pattern matrix are the beta weights of the variables. These indicate the importance of each variable in predicting the factor and take into account the correlation between the factors. Where these correlations are small the pattern and structure matrix are virtually identical. Interpretation of the structure loadings, the correlations, seems to be the more sensible since these are less likely to change in studies where different factors are used.

Other issues

The most important aspects of factor analysis have now been described. To end this chapter I shall bring together some other rather different issues, which deserve examination if the factor analysis of abilities is to be understood.

Target matrices

So far in this chapter I have implicitly, at least, been discussing exploratory analysis where the aim is to map out the field. Nevertheless on occasion exploration of this kind is not required. Rather there may be hypotheses which we wish to test, and Guilford (1967) is one of the most notable users of factor analysis for this purpose. Guilford and his colleagues, indeed, (e.g. Guilford and Hoepfner, 1971) have developed a complex description of cognitive ability by this means. Despite the computing complexity of hypothesis testing factor analysis, the principles can be described.

In this method target matrices are specified, i.e. the factor loadings of the variables are given, derived from some theoretical position. Then a Procrustes factor analysis is carried out in which, instead of aiming at simple structure, the factors are rotated to a position as close as possible to the target matrix, which can be specified in detail, with all loadings stated, or rather more loosely with high, low and zero loadings. The more loose the specification the easier the fit.

In many respects these Procrustean methods might seem to be a powerful method of hypothesis testing. Unfortunately this view can no longer be maintained. Thus Horn and Knapp (1973) showed that the Procrustes method was so powerful that hypothesis-rejection was extremely unlikely unless the target matrix was specified in great detail. They demonstrated that Procrustes rotations could match target matrices not only from random data but also from data sets with hypotheses built into them that were antithetical to the targets. From this it must be concluded that Procrustes analyses should be used only where the target matrix is specified completely.

There is a further difficulty with these analyses. If simple structure is the best solution to the matrix, as has been argued, then the Procrustes solution, by definition, is not to be chosen. Thus the best procedure would seem to be to test hypotheses by comparing these with the simple structure rotation. Where they match they are supported.

Confirmatory factor analysis

From Procrustes rotations to confirmatory factor analysis (Joreskog, 1969) is a small step, because unlike the exploratory analyses which have been discussed so far, confirmatory analysis is designed to test hypotheses. Maximum likelihood methods are used to fit target matrices which may be specified with varying degrees of precision. As Nunnally (1978) has pointed out, with large matrices the chi-square test of fit finds it difficult to choose between target matrices unless these are grossly different. In practice, therefore, subjective judgement still has to be used in selecting the best solution.

If there is sufficient computing time it would be worth while to compare the confirmatory solution with a simple structure analysis, where, of course, a clear target matrix could be specified. However, I agree with Nunnally that despite its statistical sophistication it has no real advantages compared with simple structure rotation.

This concludes all that needs to be said about factor analysis as it is generally used in the field of abilities. In summary simple structure factors are recommended because they are replicable and represent major determinants of variance. This technique, in which variables are factored, *R analysis*, is actually not the only possible method of factor analysis and to end the chapter I shall describe, briefly, some other factor analytic designs.

P technique

In this method test scores obtained from one individual are subjected to factor analysis. In practice this is difficult to carry out because a large number of variables is needed. Nevertheless the results could reveal determinants of ability within an individual.

Q technique

Here the correlations between people are factored so that factors reveal groups. Thus in a study of abilities we might find factors isolating a highly intelligent group or a group with verbal deficits, just for example. What advantages there are in this approach, compared with scoring individuals on normal R factors, are not obvious.

O technique

Here scores from two occasions from the same subject are factored. If other measures such as those concerned with education or psychotherapy are included in O analyses, some valuable factors giving insight into these processes could be obtained.

There are some other approaches but these are too speculative and specialised to deserve comment. Cattell (1978) should be consulted for further details.

Conclusions

The conclusions from this chapter are clear. I have shown that the problems caused by the infinity of solutions in factor analysis are solved by rotation to simple structure. Furthermore simple structure yields replicable factors that are important determiners of the variance. Factor analysis is, therefore, ideal for exploring the structure of complex fields. Finally it was demonstrated that there are now techniques available which will reliably produce simple structure.

In the following chapters I shall critically examine the substantive findings from the factor analysis of human abilities and ultimately the factorial picture of intelligence will emerge.

Chapter three

Factor analyses of human abilities

In this chapter I shall describe and discuss the substantive, psychological findings that have been obtained from the factor analysis of human abilities, results which may be thought of as describing the structure of abilities. As was made clear in the previous chapter, although there is a vast number of factor analytic solutions to any correlation matrix, in fact there is general agreement as to how replicable and meaningful factors may be obtained. Consequently it is pointless to discuss all the factor analytic claims concerning intelligence and the structure of abilities. Rather I shall restrict my scrutiny to those factor analyses which are technically satisfactory and whose findings, therefore, cannot be written off as statistical artefacts. This means, almost inevitably, that only work conducted or replicated within the last twenty years will be considered.

I shall first examine the work of Cattell (1971) whose list of factors is generally considered to be one of the most accurate analyses of the structure of abilities. Before I do this, however, certain of the critical issues of factor analysis, discussed in the last chapter, must be borne in mind. These are sampling of variables and sampling of subjects.

It is not possible to produce a definitive list of abilities through factor analysis, as one might produce a list of every English bird, on account of the problems of sampling both variables and subjects. In the case of variables how can we be sure that we have sampled them all? Clearly if we have not, the resulting list of factors could not be complete. Cattell attempted to overcome this problem by hypothesising a model of human ability which took into consideration the action necessary to carry out the task, the content of the test and the cognitive processes involved in it. This model guided the work of Cattell and his colleagues over the years in their factoring of abilities and provided a useful basis for sampling of variables. However, I shall not describe it in detail because, despite its value and ingenuity, it cannot yield a definitive list. This is simply because, to take one example, cognitive processes are not well enough understood to be sure that the tests encapsulate them or that some of these have not been left out, or wrongly conceived. In addition to this, and perhaps more salient, the model reveals that kinaesthetic and tactile abilities are important, yet there are few good tests of these variables. Finally in the last analysis Cattell's model, no matter how plausible, is subjective. To sample tests on the basis of it could not yield a definitive list of factors, although it would yield, without doubt, the most important factors.

Sampling of subjects is important for two reasons. First if there is little variance on a factor it is unlikely that it will emerge with proper clarity. This is important in the case of rare but clear abilities. Musical and mathematical ability may have rather little variance in some groups. Other abilities, especially of the practical kind, even if tests existed to measure them, would show little variance, in most experimental studies.

Of greater psychological significance is the fact that the structure of abilities which is so clearly found in Western society might be less clear (or even totally dissimilar) in cultures which are entirely different. This raises the problem of the status of ability factors: do they represent something about people or do they reflect aspects of the environment as it moulds our ways of thinking? This question will be answered in a later chapter of the book. Here, however, for our discussion of the structure of abilities, it is sufficient to remember that any such structure is not necessarily to be found in a different population and cannot be regarded without proper examination of the evidence, as universal.

There is one further problem involved in compiling a definitive list of factors in any area of psychology. This arises from the problem of distinguishing between a bloated specific and a group factor. It will be recalled from the previous chapter that ultimately the distinction could be made only in terms of correlations with external criteria. This usually affects only small factors which might be endlessly produced and which would require years of study before a decision could be made about their external correlations. Nevertheless it is clear that from this problem alone no list of primary factors could be definitive or complete.

Despite these caveats, the list of primary factors in Hakstian and Cattell (1974), derived from a simple structure rotation of 57 variables selected to represent a broad spectrum of abilities, represents what is generally agreed upon by modern factor analysts. The Cattell Primary Factors are set out in Table 3.1.

Table 3.1 Primary factors with brief description

V	*Verbal ability* – understanding words and ideas. Loading on synonyms, meaning of proverbs, analogies. Probably the best indicator of *gc* – crystallised intelligence
N	*Numerical factor* – this is facility in manipulating numbers which is factorially distinct from arithmetic reasoning
S	*Spatial factor* – the ability to visualise two- or three-dimensional figures when their orientation is altered
P	*Perceptual speed and accuracy factor* – which involves assessing whether pairs of stimuli are similar or different
Cs	*Speed of closure of factor* – this taps the ability to complete a gestalt when parts of the stimulus are missing. Speed of verbal closure correlated 0.61 with word fluency suggested that familiarity with words plays a part in the results
I	*Inductive reasoning* – this involves induction, reasoning from the specific to the general
Ma	*Associative or rote memory* – memory for pairs for which no mediating link exists. There are substantial correlations between the word number pairs test and figure-number pairs test although according to the Guilford model these should be orthogonal
Mk	*Mechanical ability or knowledge*
Cf	*Flexibility of closure* – this involves disregarding irrelevant stimuli in a field to find stimulus figures. According to Hakstian and Cattell (1974) this factor is a manifestation of Witkin's field independence (Witkin, 1962) and is related to the personality factor independence
Ms	*Span memory* – this is the short-term recall of digits or letters as has long been used in the WISC and WAIS tests (Wechsler, 1958). It is noteworthy that the mean correlation of the MS tests with the Ma tests (factor 7) is only 0.18
Sp	*Spelling* – recognition of misspelled words. Hakstian and Cattell (1974) point out that spelling has not appeared as a factor in previous researches because usually there was only one test, thus making the emergence of a factor impossible. Since there are good correlations with V and W whether spelling is a narrow primary or dependent on these two factors is not yet clear
E	*Aesthetic judgement* – the ability to detect the basic principles of good art. Like Mk, this would appear to depend much on previous experience
Mm	*Meaningful memory* – this involves the learning of links between pairs in which there is a meaningful link. The mean correlation of Mm with Ma tests is only 0.35, suggesting that Ma and Mm are behaviourally distinct
O1	*Originality of ideational flexibility* – this loaded on the multiple grouping tests of Guilford and Hoepfner (1971) which fall into the divergent production of semantic classes cell of the Guilford model. There are substantial correlations between this O1 factor and O2 and F1

Table 3.1 Primary factors with brief description (contd.)

FI	*Ideational fluency* – the ability to reproduce ideas rapidly on a given topic. This is distinct from WI word fluency and associational and expressional fluency which were not included in this study, although discussed by Guilford and Hoepfner (1971)
W	*Word fluency* – the rapid production of words, conforming to a letter requirement, but without meaning. This factor was found as early as 1933 by Cattell and has regularly occurred ever since
O2	*Originality* – as with O1 this is a relatively new factor loading on the Guilford tests where subjects have to combine two objects into a functional object. Originally the test was designed to mark the convergent production of semantic transformations but it actually loaded on a divergent production factor (Guilford and Hoepfner, 1971)
A	*Aiming* – involving hand–eye coordination at speed
Rd	*Representational drawing ability* – drawings of stimulus objects scored for precision of lines and curves

Source: (Kline, 1979)

Table 3.2 contains eight further primary factors which, Cattell (1971) argues from a study of previous work, would be likely to occur in correctly rotated factor analyses. Much research, it should be remembered, can be discounted simply because simple structure was not obtained.

Table 3.2 Further primary factors in ability studies

D	*Deductive reasoning*
Mc	*General motor coordination* – this is tested by the pursuit meter, among other tests
Amu	*Musical pitch and total sensitivity* – found in the Seashore (musical aptitude) test
Fe	*Expressional fluency* – found in Guilford (1967), verbal expression for assigned ideas
ams	*Motor speed* – found in Guilford (1967)
asd	*Speed of symbol discrimination* – found in Guilford (1967)
—	*Musical rhythm and timing*
J	*Judgement* – ability to solve problems where judgement and estimation play a part. Again found in Guilford (1967)

Source: (Kline, 1979)

These primary factors are correlated. Hence they can be understood fully only in conjunction with their second order analysis, i.e. when the correlations between the factors are themselves factored. The resulting five second order factors are set out in Table 3.3.

Table 3.3 Second order ability factors (Horn and Cattell, 1966)

gf	*Fluid intelligence* – loading on the Culture-Fair Test, inference, induction, memory span and flexibility of closure. Also it loads on intellectual speed and level tests
gc	*Crystallised intelligence* – this is the factor of traditional intelligence tests. Loading on verbal, mechanical, numerical and social skills factors
gr	*(now Pv) Visualisation* – loads all skills where visualisation is helpful, spatial orientation, formboards. This factor loads some of the tests of the Culture-Fair Test thus demonstrating that even here visualisation can be useful. Cattell (1971) points out that in some earlier research, visualisation had appeared as a primary but this work by Horn and Cattell clearly shows this not to be the case
gr	*Retrieval capacity or general fluency* – loading on ideational fluency, association fluency and irrelevant association test, it is the general retrieval power which accounts for a variety of skills
gs	*Cognitive speed factor* – this affects speed in a wide range of tasks although it is a minor factor in solving gf problems. This factor is speed in mechanical performance, e.g. writing or numerical computation

Source: (Kline, 1979)

Discussion of the Cattell factors

The factors in Table 3.1 are in order of size, determined by the variance for which they account. In general it is interesting to note that these factors, which must be regarded as the fundamental dimensions of human ability, are similar to those that most people have intuitively deduced from their observations of people they meet or see on television. They also fit well with the early factorial studies of Thurstone (1947).

Verbal ability is essentially verbal comprehension and skill with words. It is highly loaded on many intelligence tests, especially those which are used for selection in schools and universities. Those who are successful in the arts, journalists and writers, broadcasters, just for example, are usually high on this factor. It should not be confused with intelligence, however, since many scientists are relatively weak in verbal ability. Those high on verbal ability often strike the observer by their speed at understanding arguments and responding, often with a flood of words.

Numerical ability, *N*, is a factor of computational speed and accuracy, and is not, necessarily, implicated in high mathematical ability. It is clearly a useful factor in old-fashioned book-keeping but its practical utility in the age of computers and calculators is limited. The spatial factor, *S*, is one which has been extensively studied. It is the ability to keep orientations in mind. In everyday life we can see this factor in action in map reading and following diagrams or driving accurately, when using the mirrors. Although, as I indicated, these factors are in accord with common sense, and it would be strange if they were not since a reasonably efficient psychology is essential for the conduct of life, there is an important distinction between the spatial factor and visualisation, which is not obvious. The visualisation factor is far more broad and is, in fact, a second order factor, as will be discussed later in this chapter. A typical, salient item for this spatial factor is: if a three-inch cube is painted red and sawn into one-inch cubes, how many will have paint on one, two, three or four sides?

The perceptual speed and accuracy factor, P, is of considerable interest in the light of work on the nature of intelligence which has been done quite independently of factor analysis. This research, which will be discussed fully in Chapter 7, gives some reason to think that intelligence is correlated with the efficiency of the transmission of information through the nervous system (Eysenck, 1982). Perceptual speed and accuracy would clearly be highly related to such neural efficiency.

I shall not examine all the primary factors in Tables 3.1 and 3.2 since they are described there and constitute the best established dimensions of ability. Some comments, however, will be useful for our scrutiny of these results later in this chapter. The inductive reasoning factor is certainly central to the factor analytic notion of intelligence, as first conceptualised by Spearman, as reasoning ability. It is the fact that most problems in any field require reasoning for their solution that makes intelligence so general a factor. It cannot be sufficiently emphasised that high intelligence is not the same as massive learning. They may go together since intelligence makes the accumulation of knowledge easier. But the correlation is far from perfect. The intelligent person is far more easily spotted from his response to *new* problems not his knowledge of old solutions.

Notice that there are two fluency and two originality factors. These and similar factors have been extensively studied over the years by Guilford and his colleagues (e.g. Guilford and Hoepfner, 1971), work which we shall examine later in this chapter. Here it is important to point out that one of Guilford's criticisms of the standard factorial conception of intelligence was that it exaggerated the importance of logical thinking at the expense of creativity, these fluency and originality factors being highly loaded on creative thinking. This difference between these two modes of thought is often indicated by the terms convergent and divergent thinking. These have been studied in connection with originality in the sciences and arts by Hudson (1966) who, not unexpectedly, found scientists, in general, to be convergent and conforming.

Mention of creativity brings me on to the factors in Table 3.2. Here we find two musical factors. However, this is almost certainly not the total number of factors in musical ability, as the survey by Shuter (1968) demonstrates. There are efficient tests of musical ability, for example those by Wing (1936) and Seashore (1919), but their factorial structure is not clear. This illustrates the problem of attempting to provide a complete list and escaping the dangers of bloated specifics. Certainly Shuter's argument that there must be a general musical factor, based upon unrotated components, will not do since principal components analysis always produces such a first general factor, as a simple artefact of the algebra.

Factor analysis, it was claimed in the previous chapter, simplified a complex set of data. Yet, in the case of these primary factors, 27 in number and far from complete, this can hardly be said to be so. In addition it is noticeable that intelligence has not so far emerged. Clearly some method of organising these primary factors is required.

Since these factors have been rotated to the oblique position in the search for simple structure, the obvious approach to simplification is to factor the intercorrelations between the primaries and examine the second order factors and, if necessary, the higher order factors. This has been done and the second order factors are set out in Table 3.3 from Horn and Cattell (1966). The implication of this table is that human abilities may be conceived in terms of five capacities – fluid and crystallised intelligence, visualisation, fluency and cognitive speed. These five factors embrace most of the variance in human ability and here there can be no doubt that factor analysis has simplified the field. Before examining the nature of these five factors in a little more detail, it must be emphasised again that auditory, tactile and kinaesthetic skills have not been included in these analyses so that this list of second order factors is not complete. Nevertheless, despite that caveat, these factors are of considerable psychological significance.

The first point to note is that intelligence has split into two factors.

Since intelligence is the central issue of this book, I shall leave discussion of these factors until I have examined the other three. As might be expected visualisation, the third factor, loads all those tests and skills in which the ability to visualise is important. Many engineering, architectural and geometric problems involve this factor, for example, and it may well be the one that distinguishes the practical person from those who are hopeless at anything with their hands. Fluency or retrieval capacity is an interesting and important factor. This refers to the ability quickly to access material in your memory. For example, the tests that load on it are association fluency and ideational fluency where subjects are required to think up uses for objects. Let us take coal as an example: light a fire, make marks, build a wall, line a drainage pit, use as a missile, use as a ball, make bootpolish, dye clothes, use as eye make-up and so on. Now most of us can immediately understand all these answers even if we could not have thought of them and some will have thought of many other responses. This fluency is most impressive and can be easily observed in good speakers and writers. This is one of the factors that Guilford has concentrated upon in his work on divergent abilities and it appears to be implicated in creativity. Certainly the majority of creative individuals, both writers and musicians, are fluent: Haydn, Telemann, Dickens and Scott are obvious examples of incredible fluency. It has been said that it would be difficult simply to write down all the music Telemann wrote in a lifetime, let alone compose it. Yet in addition to this he prepared the plates for printing it. Telemann was clearly extremely high on *gr* and also on the last factor, the cognitive speed factor which involves speed of doing anything, e.g. writing numerical computation. When this is combined with high intelligence and fluency it is likely that a truly impressive person would emerge. Notice, finally, to continue with our musical example, that these five factors must be incomplete since there is no musical factor in this list.

This brings us to the two large second order factors, crystallised and fluid ability, which have been extensively studied by Cattell and colleagues (Cattell, 1967, 1971). Fluid ability is a basic reasoning ability which can be applied to a wide variety of problems, hence its emergence as a broad second order factor. Fluid ability is close to the *g* of Spearman, the ability to educe correlates, and, as we shall see in a later chapter of this book, is considerably genetically determined. However, it is less than *g*, because essentially in this modern factor analytic work, in which factors are rotated to simple structure, the old *g* has split into two. This second factor is crystallised ability.

Crystallised ability is the factor of traditional intelligence tests, especially those which were largely verbal. It consists of fluid ability as it is evinced in the skills valued by a culture. Thus in Great Britain and the USA fluid ability is implicated in technological and verbal skills as taught in school and university. However, in a culture far different from that of the West, for example that of the Aborigines, or the Kalahari Bushmen, fluid ability is likely to be evinced in skills far different. These problems of comparing abilities cross-culturally will be fully discussed in a later chapter.

The relation of these two intelligence factors is highly revealing and it has been explicated in considerable detail both theoretically and empirically by Cattell (1971) under the name of investment theory, which can be briefly summarised. In this theory it is claimed that there develops a single, general relation-perceiving ability which is connected to the total associational neuronal development of the cortex. This is fluid ability, which is obviously highly heritable. Crystallised ability, on the other hand, develops as a result of investing fluid ability in particular learning experiences. Thus at an early age, say 2 or 3 years, fluid ability and crystallised ability are highly correlated. As children grow older and undergo different experiences at school and in the family, so, clearly, fluid ability and crystallised abilty become less highly correlated. The bright and well-adjusted child who attends a good school and receives encouragement at home will invest most of her fluid ability in the crystallised skills of her culture. On the other hand, the equally bright child from a home where education is not valued and who attends a school of indifferent quality will not thus invest his fluid ability. His school performance may be far worse than a moderate child who invests all his ability at school.

This dual nature of intelligence accounts for many of the disparate claims that have been made concerning the nature of intelligence, claims which will be examined in detail in later chapters of this book. In general where studies indicate that environmental factors play an important role in the development of intelligence it is likely that the intelligence tests were largely measuring crystallised ability. Where, on the other hand, the influence of the environment appears to be relatively trivial, the tests are probably loaded on fluid intelligence.

There is one final point concerning the work of Cattell which deserves mention. This is the third order analysis of the factors. As Cattell (1971) points out, there are considerable technical problems in carrying out third order analyses such that few have been attempted.

Horn and Cattell (1966) report one such research although the findings must be regarded as tentative. A factor emerged at the third order which loaded on fluid and crystallised ability among children aged 5 to 14 years. Among adult criminals, however, it failed to load crystallised ability but also loaded the other second orders. Cattell (1971) accounts for these anomalous findings by arguing that in the young sample the second order retrieval, visualisation and perceptual speed are not clear cut so that this third order factor could not load on them. As regards the failure to load the crystallised factor in the criminals this was due to the fact that criminals do not invest their fluid ability in the skills valued by the culture and he cites work where there is a large discrepancy in the fluid and crystallised ability scores of criminals (Warburton, 1965).

If we ignore these discrepancies how is this factor to be interpreted? It cannot be the original Spearman factor because its main loading is on fluid ability rather than the crystallised ability of traditional intelligence tests. In fact Cattell labels it historic fluid ability, the childhood fluid ability which gave rise to the current fluid ability, a name which should be regarded as highly tentative.

Recently Cronbach (1984) has argued that fluid ability is probably the old *g* factor, an argument supported by the reanalysis of Horn and Cattell's work by Undheim (1981), although there are complex technical difficulties associated with his claims. Gustaffson (1988) makes the point that this problem of the identification of fluid ability and Spearman's *g* is simply a technical one which can be solved by maximum-likelihood model-fitting methods (which were discussed in our section on confirmatory factor analysis). He reports research with adolescents where fluid ability is clearly to be identified with the old *g*. This means that crystallised ability resembles Vernon's old V.Ed factor (Vernon, 1961), evinced in scholastic attainment. Clearly more research with confirmatory analysis is required but his solution, which is not radically different from that of Cattell, is psychologically meaningful. Clearly it makes good sense to measure both fluid and crystallised ability if we want to use tests as predictors in real-life settings.

Conclusions from Cattell's work on the structure of abilities

As I indicated at the beginning of the chapter, I want only to explicate the structure of human abilities and the place of intelligence within that structure, in technically competent factor analyses. I shall not delineate the nature of intelligence, as it thus emerges, because this entails the consideration of complex lines of evidence which are dealt with best in separate chapters. Nevertheless some summary conclusions from the extensive factor analytic work by Cattell and his colleagues must be drawn.

It is not correct, as Spearman claimed, that human ability should be conceptualised in terms of one general factor, intelligence, together with a number of specifics. This picture is too simple. Rather we should see human ability as encapsulated by five factors, the second orders which have been described. Of these, however, two highly correlated factors are by far the most important, fluid and crystallised ability, the latter the cultural emanation of the former. Fluid ability is essentially a basic reasoning ability, necessary for problem solution on a wide variety of problems and highly heritable, being dependent on neuronal efficiency. These two intelligence factors are Spearman's *g*, split by more efficient factor analysis. Thus, in summary, Cattell's delineation of the structure of ability is an amplification of that of Spearman. It is instructive to remember here that Cattell was a student of Burt who, himself, followed Spearman into the Chair at University College London. The more modern work of Undheim and Gustaffson does little to change this picture where we see one broad factor, fluid ability or Spearman's *g*, and four group factors, as described by Cattell, but less general than fluid ability.

Discussion of the Guilford model

I shall now scrutinise the work of Guilford, who over many years has investigated the structure of human abilities and is rightly regarded as one of the major researchers in this field, his most important book being *The Nature of Human Intelligence* (Guilford, 1967). I intend to discuss the basic Guilford model, known as the Structure of Intellect Model, which has a number of problems, so severe that more modern attempts to enlarge it do not appear worthwhile. Despite these flaws this model must be discussed because, beyond the field of psychometrics, it is still considered to be an important description of human abilities.

Guilford and his colleagues differ from the majority of workers in the factor analysis of abilities on two important counts. First they argue that simple structure is best attained with orthogonal factors and that the apparently improved simplicity of oblique factors is more than compensated by the complexity of the factors being correlated. One considerable disadvantage of orthogonal factors is that analysis can go no further: no higher order factors are possible. Guilford (1967), however, overcomes this by estimating the factor correlations from the factor loadings. However, if the factors are correlated the oblique position is best. That Guilford is prepared to do this underlines the incoherence of this approach.

His second point of difference lies in the fact that he denies the positive manifold of the ability sphere, claiming that not all human abilities are positively correlated (Guilford, 1964). Indeed he argued that about 20 per cent of the correlations which he had obtained between tests of ability in his laboratory were not significantly different from zero. From this finding a profound difference between Guilford and the majority of factorists arises. He does not support the postulation of a *g* factor or factors.

In brief his claims are antithetical to those of Cattell – no *g* factor, in any guise, and a large number of orthogonal factors. Nevertheless, as was mentioned above, there are good reasons for rejecting the Structure of Intellect Model and these are set out below.

Description of the model

The model, which represents a method of classifying tests, states that there are 120 independent abilities, each characterised by an intersection of one of five mental operations (cognition, memory, divergent thinking, convergent thinking and evaluation) on one of four contents (figural, semantic, symbolic and behavioural) to produce one of six products (units, classes, relations systems, changes and implications). This is, it can be seen, a three-dimensional model with 5 x 4 x 6 intersecting cells, each representing an independent ability, hence the 120 factors. This Structure of Intellect Model is supported by numerous factor analyses which have been reported in Guilford (1967) and Guilford and Hoepfner (1971).

The first objection to this model arises from the fact that the category of operations is arbitrary and not inclusive. The weakness here is that these operations are not derived from any theory or from empirical study but are simply intuitive categories in Guilford's head, as it were. This is not a good basis for a model. Another objection can be launched to the category of products which, again, is not comprehensive and which could be abandoned with little loss, as Butcher (1973) has argued.

These problems illustrate the more general criticism of this model, raised by both Eysenck (1967) and Cattell (1971), namely that it is divorced from psychological theories of cognitive processing or human ability although it was used as a basis to construct tests. In later sections of this book I shall examine research into intelligence which is directly linked to cognitive psychology.

The arbitrary nature of this model has been mentioned because it has been used as a basis to construct tests and the results from these tests are then claimed to support the model. A few examples will illustrate this circularity. Guilford (1964), for example, makes it clear that his objective is to produce tests such that they each measure only one of the 120 abilities, i.e. load on one of the primary factors, when they are rotated orthogonally. This, of course, is natural, if he wants to fit the model, but it must be stressed that the model is arbitrary. This procedure minimises the correlations between any pair of tests and thus minimises any *g* loadings.

In addition to this the samples in Guilford's studies were drawn from those undergoing officer training and were, therefore, all above average intelligence. This homogeneity on the *g* factor would further contribute to small loadings. Thus it can be seen that the manner of test construction, attempting to devise tests with minimal correlations between them to fit the model, and using a sample already selected for intelligence, are bound to reduce the influence of *g*.

This brings us to a further more general point which has been mentioned in the previous chapter on factor analysis, namely the difference between a group factor and a bloated specific. There it was pointed out that by duplicating items a specific factor could be enlarged so that it resembled a group factor. Now this is essentially what has happened in the structure of intellect model. It is inevitable, if one thinks how any test constructor would go about trying to produce tests, that he tries to fill out each cell. As Eysenck (1967) has also argued, it is possible to go on subdividing items and forming factors which have no relation to anything in the real world or to any human ability. This is the danger in the crude use of factor analysis in the social sciences.

Guilford and Hoepfner (1971) have argued that the structure of intellect model is supported by their factor analyses in which 98 independent factors had been isolated. Thus their counter to all the objections to the model is simply that, in the end, the factor analyses support it. However, as I indicated in the previous chapter, when discussing rotation to target matrices, it was shown by Horn and Knapp (1973) that the Procrustes rotations, used by Guilford, were incapable of rejecting hypotheses and thus could not be regarded as support for any model, especially where, as in this work, there was over-factoring. This destroys much of the empirical confirmation for Guilford's work.

An even more severe difficulty in testing this theory has been noted by Undheim and Horn (1977). To carry out one large factor analytic study of the 120 factors would require 480 tests to determine the factors clearly. Since they claim that five times the number of subjects to variables is necessary, 2,400 subjects are required. If each test were only 5 minutes in duration this would require 96,000 hours of testing. If we halved the subjects this is still an enormous task which has not been carried out.

It might be thought more simple to study segments of the matrix separately. However, since there are 7,140 paired comparisons among 120 factors all of which must be made to demonstrate factorial independence this is even more time consuming. If 18 factors could be examined in each factor analysis 364 studies would have to be made demanding, each one, 6 hours of testing and 360, or at least 130, subjects. Thus to obtain full support for the theory is a truly gargantuan task, if factor analysis is to be used.

Conclusions concerning the Guilford model

All these arguments show clearly that there is no strong evidence in support of the Structure of Intellect Model of human abilities. At best the model is a useful rough guide to test construction but it is rough because the provenance of this model is human intuition rather than empirical evidence or psychological theory.

Other work in factor analysis

Two factor analytic sets of results have now been examined. Those of Cattell represent an extensive and technically sound list of factors which have received broad acceptance in psychometrics. Those of Guilford have been shown to be defective and his model is of more interest for test construction than as a model of human abilities. However, it would be wrong to create the impression that this completes the factor analytic picture. Some other work must be mentioned although, as we shall see, there are essentially few differences from Cattell's description.

Ekstrom, French and Harmon

First I shall describe the work of Ekstrom, French and Harman (1976) which is the culmination of the search for ability test factors which has been going on for many years (e.g. French, 1951). Essentially this kit of tests consists of the tests which best identify factors in the field of ability which, from a search of the literature, regularly emerge. Their list of factors, therefore, ought to represent, as does the list of Cattell, the factor analytic description of human abilities. Indeed, based as it is on a search of all relevant studies, the picture thus provided ought to be definitive, but should, nevertheless, differ little from that of Cattell, since that description was claimed to be replicable, having eliminated technical problems. The 23 factors are set out in Table 3.4.

Table 3.4 Ekstrom factors

CF Closure, Flexibility of	MV Memory, Visual
1 Hidden Figures Test	1 Shape Memory Test
2 Hidden Patterns Test	2 Building Memory
3 Copying Test	3 Map Memory
CS Closure, Speed of	N Number
1 Gestalt Completion Test	1 Addition Test
2 Concealed Words Test	2 Division Test
3 Snowy Pictures	3 Subtraction and Multiplication Test
	4 Addition and Subtraction Correction
CV Closure, Verbal	P Perceptual Speed
1 Scrambled Words	1 Finding A's Test
2 Hidden Words	2 Number Comparison Test
3 Incomplete Words	3 Identical Pictures Test
FA Fluency, Associational	RG Reasoning, General
1 Controlled Associations Test	1 Arithmetic Aptitude Test
2 Opposites Test	2 Mathematics Aptitude Test
3 Figures of Speech	3 Necessary Arithmetic Operations Test
FE Fluency, Expressional	RL Reasoning, Logical
1 Making Sentences	1 Nonsense Syllogisms Test
2 Arranging Words	2 Diagramming Relationships
3 Rewriting	3 Inference Test
	4 Deciphering Languages

Table 3.4 Ekstrom factors (contd.)

FF Fluency, Figural
1 Ornamentation Test
2 Elaboration Test
3 Symbols Test

FI Fluency, Ideational
1 Topics Test
2 Theme Test
3 Thing Categories Test

FW Fluency, Word
1 Word Endings Test
2 Word Beginnings Test
3 Word Beginnings and Endings Test

I Induction
1 Letter Sets Test
2 Locations Test
3 Figure Classification

IP Integrative Processes
1 Calendar Test
2 Following Directions

MA Memory, Associative
1 Picture-Number Test
2 Object-Number Test
3 First and Last Names Test

MS Memory Span
1 Auditory Number Span Test
2 Visual Number Span Test
3 Auditory Letter Span Test

S Spatial Orientation
1 Card Rotations Test
2 Cube Comparisons Test

SS Spatial Scanning
1 Maze Tracing Speed Test
2 Choosing A Path
3 Map Planning Test

V Verbal Comprehension
1 Vocabulary Test I
2 Vocabulary Test II
3 Extended Range Vocabulary Test
4 Advanced Vocabulary Test I
5 Advanced Vocabulary Test II

VZ Visualisation
1 Form Board Test
2 Paper Folding Test
3 Surface Development Test

XF Flexibility, Figural
1 Toothpicks Test
2 Planning Patterns
3 Storage Test

XU Flexibility of Use
1 Combining Objects
2 Substitute Uses
3 Making Groups
4 Different Uses

Source: (Kline, 1979)

These are primary factors and it is, therefore, instructive to compare them with the list in Table 3.1. It is immediately clear that there is considerable overlap. The differences are due to the fact that the Ekstrom list is derived from a literature search regardless of the technical problems associated with the research. Thus there are more fluency factors and this reflects the influence of Guilford and his colleagues, whose factors, as we have seen, do not stand scrutiny. Nevertheless what is important about the list is that the main Cattellian primaries are there: induction, number, perceptual speed, visualisation, logical reasoning, spatial reasoning and verbal comprehension. There can be no doubt that the Ekstrom list confirms the Cattellian factors and that any differences are only those of emphasis and result from differing factor analytic methods.

Because the kit is primarily concerned with tests, rather than factor structure, there is no indication of the intercorrelations of these factors or of the second order structure. However, since these factors are little different from those in the Cattell list, there is no reason to think that the second order factors would be different. Certainly it is easy to discern fluid and crystallised intelligence, visualisation, retrieval and memory. In conclusion I think that it is fair to argue that the factor analytic picture provided by Cattell and Ekstrom is essentially and psychologically identical.

Carroll

Next I shall examine the work of Carroll. Carroll, who is one of the most thorough and technically proficient scholars in the field of the factor analysis of abilities, has over the years attempted to integrate all factor analytic researches. What is remarkable about this effort is that wherever possible he has obtained the original data and refactored it according to the technical criteria which were discussed in Chapter 2 so that simple structure was obtained. Although the results of this Herculean task are not yet available, an interim report (Carroll, 1983) gives some indication of the findings.

Because Carroll has surveyed so many studies, the list of primary factors is enormous, as it is bound to be. Remember that, at the beginning of this chapter, it was argued that a complete list of factors is, almost by definition, impossible. However, the list of second orders is highly revealing. Almost all studies support the Cattell list: fluid intelligence, crystallised intelligence, general visual perception, general memory and fluency. Indeed the only real point of difference at the second order between Cattell and Carroll is that the speed factor does not seem as clear in Carroll's list, while he finds an auditory perception factor, of which, of course, Cattell had no tests.

Gardner

Finally I shall briefly mention the work of Gardner (1983) which is not factor analytic at all but which, based upon the study of exceptionally talented individuals, sets out a structural theory of intelligence. For Gardner the importance of *g* is hugely inflated. In its place he suggests multiple intelligences: musical intelligence; bodily-kinaesthetic intelligence; linguistic intelligence; spatial intelligence; interpersonal intelligence and intrapersonal intelligence (as displayed by Virginia Woolf). I find nothing striking about this theory or antithetical to the normal picture of the structure of abilities. Most of these factors are well-established primaries, correlated, of course, whose structure is better understood in terms of the *g* factors as has been discussed. Of course in some individuals, special primary factors are more influential than the broader *g* factor. All analyses of intelligent performance take due account of these special factors when individuals score highly on them. This work illustrates the danger of extrapolating from special populations (the gifted) to normals. In the former special factors are all important. In the latter the general factor assumes greater significance. Actually to cite Virginia Woolf as an exponent of intrapersonal skills is perhaps questionable, given that she committed suicide.

Conclusions

There is little doubt that, contrary to the critics of factor analysis who are ignorant of the technique, there is good agreement about the structure of human abilities. Essentially there are two intelligence factors, highly correlated, one dependent on neural structure, the other its cultural emanation, both implicated in most problem solving. However, the recent work by Undheim (1981) and Gustaffson (1988) which claims that fluid ability is essentially Spearmen's *g* should not be forgotten. In addition to these two factors, perceptual speed and accuracy, fluency, visualisation and cognitive speed are also important. The remainder of this book will be concerned with the psychological meaning, measurement and implications of these ability factors, concentrating, in the main, on the two intelligence factors.

Chapter four

Measuring intelligence

There are more tests of intelligence than any other psychological variable and intelligence is tested in a wide variety of applied contexts, often where the results are of prime significance for subjects, as in selection for jobs or educational courses. Indeed, at one time, as has been made clear by Kamin (1974), intelligence testing was used to keep out certain national groups from the USA. Similarly, in Great Britain, for many years after the Second World War entry to grammar school was dependent upon scores on intelligence tests, except, of course, for those who could pay. For these reasons alone there have been many objections to intelligence tests of which the most common is that they measure only what intelligence tests measure. This claim will be systemmatically destroyed in later sections of this book by simple empirical research. However, in this chapter I intend to examine some of the most widely used intelligence tests because a proper understanding of their nature, at the item level, will be valuable for understanding some of the arguments used later in this book.

Before we can begin our assessment of intelligence tests a few simple but important terms which will be frequently used in our discussions must be defined.

Definition of terms in intelligence testing

Reliability

Test reliability has two different meanings. One refers to reliability over time. If a subject takes a test on two occasions, given that his status on the variable has not changed, he should obtain the same score on each occasion. The correlation between scores on different occasions indicates the reliability of the test. Ideally it should be 1, perfect, but in an imperfect world this is rarely, if ever, so. However, if a test is to be used with individuals it should certainly be greater than 0.7. The other meaning of reliability refers to the internal consistency of the test, to what extent the items are intercorrelated. High reliability is clearly important, in this sense, for if the items are not measuring the same variable, there must be something wrong with the test. However, by having items that are highly similar, reliability can be raised but at the expense of validity. Such a test measures only what the test measures and is thus, in factor analytic terms, a bloated specific. For these reasons very high internal consistency reliability should be regarded with some suspicion.

Validity

A test is said to be valid if it measures what it claims to measure. Obvious though this definition is, it is often difficult to demonstrate that a test is valid and there can be no doubt that a large majority of published psychological tests are not valid, as is well illustrated by a study of the *Mental Measurement Yearbooks* (e.g. Buros, 1978), in which objective assessments of tests are published.

As Vernon (1961) has stressed, there is no one validity coefficient for a test because a test is valid for some purpose – selection, guidance or clinical discrimination, just for example. In fact there are three important approaches to the assessment of validity.

Concurrent validity

The concurrent validity of a test is indicated by its correlation with another, similar test taken at the same time. Thus if I had constructed an intelligence test I could validate it by correlating it with another well-established test. There are two problems here. With many psychological variables there is no established test so that a concurrent validity study is impossible. If there is a test, on the other hand, sufficiently valid to be used as a criterion, there seems little point in developing a new one. In the case of intelligence, there are two benchmark, individual tests which I shall discuss below. However, there is some point in trying to develop a new test because these are lengthy to administer and require some skill. A brief 5 minute test which had high concurrent validity would be extremely useful.

Predictive validity

The predictive validity of a test is indicated by its correlation with some criterion in the future, hence the name. If it can be demonstrated, such a correlation is powerful support for the validity of a test. As it happens the intelligence test is highly suited to the demonstration of predictive validity. Thus if scores on an intelligence test at 11 years of age can predict university performance or promotion within learned professions, especially where control groups of similar social class but lower intelligence are also studied, this is good evidence of test validity. However, for many other psychological variables it is not clear how predictive validity could be demonstrated.

Construct validity

The construct validity of a test is a more complex matter. Here we think of all the possible hypotheses concerning the variable and put these to the test. If all or the majority of them are supported then the test is valid. An example will clarify the point. From our knowledge of the construct or concept of intelligence the following hypotheses could be drawn up.

1. Scores on the test should correlate with scores on other intelligence tests but should not correlate with scores on other tests.
2. Scores should predict performance at school, university, and job success.
3. There should be mean differences between the scores on the test of different occupational groups.

Clearly if all these hypotheses were supported it would be difficult to argue that the test did not measure intelligence. Unfortunately in the real world, there is rarely a clear-cut demonstration and the problem with construct validity is that, in the end, it is reliant on subjective judgement. For a fuller discussion of all these problems see Kline (1986).

Face validity

Finally I shall mention face validity which refers to the appearance of a test. If it looks valid a test has face validity. In fact there is little relation between face validity and true validity but it is important to retain the interest and co-operation of adult subjects. If they think they are wasting their time on a pointless exercise, then results are worthless. Children, who are more used to being told what to do, may not resent it so much if tests appear ridiculous or pointless, but group testing can lead to problems with tests of low face validity.

In brief it is desirable to have tests of high reliability and validity and with reasonable face validity if they are to be used with adults. One final psychometric term now demands definition – norms.

Norms

Norms are sets of scores from samples of different populations. They are important only in so far as they enable the test scores to be interpreted by comparison with the norms. If the norms are based on large (5,000 or more) and representative samples, then a score from any individual can be interpreted with some confidence. For example if we know that a score of 149 on a particular intelligence test is in the top 0.5 per cent of scores, then we can be sure that all subjects obtaining this score are highly intelligent.

IQ: the intelligence quotient

The IQ is one piece of psychological jargon which has percolated through to general usage. Originally it was a quotient derived from norms which were stated in terms of mental ages. Thus if a child obtained the score of an average 8-year-old his mental age, according to that test, was 8 years. However, this was clearly a very different performance if the child concerned was 10 or if she was 5. Thus the intelligence quotient was developed. This was mental age divided by chronological age multiplied by 100. Our three examples will exemplify this type of IQ. The 10-year-old has an IQ of 80 using the formula. The 5-year-old has an IQ of 160 while the average 8-year-old with a mental age of 8 has an IQ of 100. This is the source of the average IQ being 100.

However, there are many problems with this method of calculating intelligence quotients, the most severe being that there is no guarantee that a three-year difference at the age of say 5 is the same as a three-year difference at 15. The scale, in other words, has unequal intervals. The IQ now is based upon deviations from the mean and the properties of the normal bell-shaped curve, the Gaussian curve, which is set out in Figure 4.1.

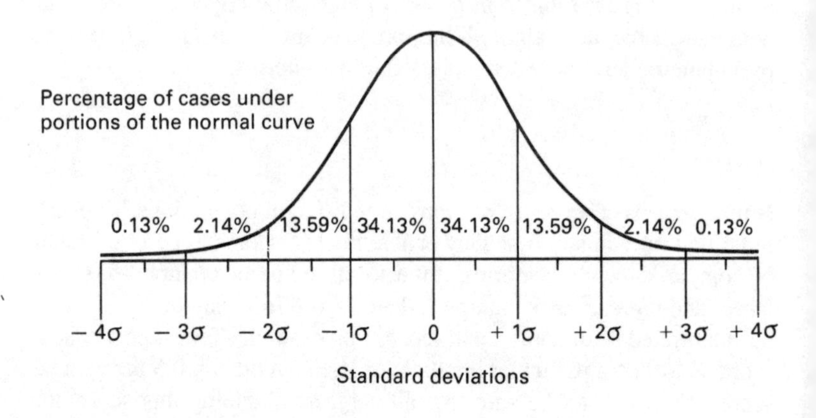

Figure 4.1 Bell-shaped normal distribution
Source: Adapted from Jensen (1980)

It is assumed that intelligence has a normal distribution as is the case with most polygenetic variables. The mathematical properties of the normal distribution have been fully explored and full details may be found in any statistical text (e.g. Nunnally, 1978). What is important here is that the area under the curve between any two points can be simply computed. Thus in a normal distribution approximately 68 per cent of all scores fall between the mean and one standard deviation, the mean being the average of the scores in the distribution and the standard deviation a measure of their scatter or variance. We have discussed variance in our chapter on factor analysis and it should be noted that the standard deviation is the square root of the variance. Approximately 95 per cent of all scores fall between the mean and two standard deviations and 99.9 per cent between the mean and three standard deviations.

The importance of all this for understanding the IQ can now be seen. Norms for intelligence tests are set up such that the average is 100 and the standard deviation 15 in a normal distribution. If we want IQs for children we standardise each age group to these norms. This means that any score of 100 always has the same meaning: it is the average for that individual's age group. Similarly any score of 130 is always two standard deviations above the mean and a score of 70 two standard deviations below it. Thus any one score represents an identical performance relative to the standardisation sample from whom the norms were obtained; 100 was selected as the mean to make these deviation norms roughly equivalent to the old procedure. An IQ of 130 can now be seen to be a high performance, putting the individual in the top 2.5 per cent of the population, the rest of the 5 per cent being below 70.

Intelligence testing

Before I scrutinise intelligence tests one further general point needs to be made. Many of these measures, including the most famous and widely used, were constructed before the structure of abilities was elucidated. That they measure intelligence is due to the psychological insight of their makers and to the pervasive character of *g*, the general ability factor, such that almost all cognitive tests load upon it to some extent. However, as we have seen, modern factor analysis has split *g* and thus it must come as little surprise that many older intelligence tests are not factor pure but measure a mixture of crystallised and fluid ability.

There are two groups of intelligence tests: group tests with which, as the name implies, large numbers can be tested simultaneously, and individual tests which are used where there is some uncertainty about an individual and where it is necessary to see how a person tries to solve the items.

Individual intelligence tests

Two individual intelligence tests stand out as benchmarks in the measurement of intelligence, although both were constructed before the structure of abilities was clearly defined. These are the Wechsler scales and the Stanford-Binet test.

The Wechsler scales

The first of these was the Wechsler Bellevue Scale of 1938. Since then a new adult scale, the WAIS of 1955, and children's scales, the WISC and the WPSSI, have been produced. It is difficult to give the best references for these tests since they are constantly under revision but the latest editions and manuals are available from the Psychological

Corporation, New York. The publications by Wechsler which contain the most useful information are Wechsler (1944, 1958, 1974), together with the latest manuals. In all they allow the individual testing of children of 4 years of age up to adults more than 70 years old.

The adult test contains eleven sub-scales, divided into performance and verbal tests, which yield a profile of scores, as will be seen below. This profile looks tempting to interpret but is fraught with danger because each scale is, on account of its length, of lowered reliability. The verbal tests are: information, comprehension, arithmetic, similarities, digit span and vocabulary. The performance tests are: digit symbol, picture completion, block design, picture arrangement and object assembly.

Verbal tests

In order not to destroy the confidentiality of the test I shall discuss items similar to those in the scales but not those in the actual test. I shall restrict my comments to the original adult scale, the WAIS.

Information This is essentially a test of general knowledge. Typical items concern capitals and geographic locations as well as authorships and the contents of parts of the Bible. Clearly this is a test of crystallised ability and is likely to be influenced by social class and education. The rationale for its inclusion in a test of intelligence is that, given equal opportunities, the intelligent person learns more quickly than the unintelligent because she can see more connections between new material and what she already knows.

Comprehension This test measures reasoning ability by requiring subjects to explain the meaning of sayings, such as a stitch in time saves nine, and to solve practical problems, e.g. finding your way when lost. However, there is some burden of knowledge if the questions are to be answered correctly. For example, taxes and prices of land enter into questions. Nevertheless this is a good test of reasoning, the ability to penetrate to the heart of the matter, and it would certainly be expected to load highly on crystallised ability.

Arithmetic This tests simple arithmetic problem solving of the traditional kind: if 20 men took two years to build a wall how many would 50 take? It must be highly dependent on learning and experience but nevertheless would be expected to load on crystallised ability, on account of the problem solving required.

Similarities In similarities subjects are required to say what is similar about two things, for example a play and a sonata. The correct answer demands that the essential rather than the superficial similarity is mentioned. In this case the response that they are both created would be scored correct, but the answer that they are both posh would not. Again this is a test that would load crystallised rather than fluid ability.

Digit span Here subjects are read digits at a steady rate and are required to repeat them, forwards and backwards. This is clearly not a test of crystallised ability and that it loads on fluid ability at all, as it does, has considerable theoretical implications for the nature of intelligence, which will be discussed in later chapters. It is interesting to note that, as Jensen (1980) points out, the backward digit span has a higher *g* loading than the forward span. This is because it requires considerably more mental processing and is more complex, the two characteristics of tasks loading high on *g*, both fluid and crystallised.

Vocabulary This is a test which many practical psychologists would use if they were forced to estimate intelligence with one brief test. As the work of Cattell (1971) shows, it loads highly on fluid ability at an early age but gradually shifts over to crystallised ability for obvious reasons. Later on in life it is argued that vocabulary scores reflect the old childhood fluid ability rather than actual fluid ability since subjects with a high vocabulary may be quite unable to deal with a novel problem (Jensen, 1980). Vocabulary loads on intelligence for the same reasons that information does so – the intelligent person's greater ability to make connections and thus remember well.

These five verbal tests show clearly that if the WAIS does measure intelligence this part of it must measure crystallised intelligence. Almost all these tests deal with intelligence as it is manifested in our culture. There can be no doubt, from an examination of these items, that the verbal WAIS scores are influenced by social class and education. That they can make good educational predictions is hardly surprising since learning plays so important a part. Nevertheless the scores are not just dependent on educational experience. They do demand reasoning and mental processing, or in the case of vocabulary, the scores reflect previous such activity.

Performance tests

One of the great strengths of the WAIS is that in addition to the verbal scales it has non-verbal or performance scales and these are far less likely to be affected by educational and social advantages. Before I examine the items in these scales I want to remind readers of Cattell's claim that to test fluid ability it is necessary, as far as is possible, to use items that are equally unfamiliar to all subjects, thus negating social factors, or so familiar that everyone in a culture has been exposed completely to them. These latter items have been found extremely difficult to construct and it may literally be impossible so to do.

Digit symbol In this test symbols and digits are coupled. With the examples before them subjects have to fill in the correct symbols for the digits. It is a speeded test. This is a test which requires relatively simple processing but considerable speed. Few subjects are unable to do it. It is, therefore, likely to be a test of the cognitive speed factor and to load only moderately on *g*, probably fluid intelligence. It is, however, easy to see why it should correlate reasonably well with success in school and in occupations.

Picture completion Here subjects are given 20 seconds to say what is missing from a picture. In all there are 21 items with such details as water in a jug and the threads of screws missing. This is obviously a task which requires some familiarity with pictures such that the denizens of some cultures might be unable to do it (Deregowski, 1980). Nevertheless there will be few adults so deprived as to lack all such pictorial experience in Great Britain. It would be expected to load on fluid ability and perhaps the perceptual factor.

Block design In block design subjects are presented with patterns that have to be made by putting blocks together. This is a test that would be expected to load on fluid ability for it is not a regularly practised skill and on the spatial and perceptual factors. I should add that it is possible that there is a trick in this test. I watched one subject divide the picture of the blocks into rectangles which greatly simplifies the task.

Picture arrangement In this test each item consists of a sequence of pictures which when assembled in the correct order tells a story. The task is to arrange them into their correct orders. As with all the performance tests most subjects will have had little prior experience of such tasks and consequently this scale should load on fluid rather than crystallised ability.

Object assembly This test resembles a jigsaw in that pieces have to be assembled to form their correct shapes as quickly as possible. Again a test of spatial and fluid ability.

Such is the WAIS, the most widely used individual test of intelligence, still regarded, despite its prefactorial design, as a criterion or benchmark for intelligence tests, although the scores must be influenced by educational opportunities and experience and thus by social class. This objection, however, is not quite as devastating as first appears for two reasons. First it is a fact, however unjust, that crystallised intelligence does depend upon opportunities to invest fluid ability. Where actual rather than potential performance is required crystallised intelligence may be a better guide than fluid ability. The second reason is that if the performance IQ is well above the verbal IQ then there is every reason to suspect that the subject may have not had good educational opportunities and hence may be more able than the overall IQ might suggest. Certainly highly educated middle-class subjects would be expected to do better on the verbal than the performance tests (e.g. Jensen, 1980).

Cattell and Johnson (1986), who support the Cattellian view of the structure of abilities, have examined some of the best known intelligence tests to ascertain their factorial structure. As is suggested from the nature of the test, essentially the verbal tests measure crystallised ability and the performance tests fluid ability, although the digit span test does not load highly on either. This, of course, measures the second order memory factor which was discussed in Chapter 3. Cattell and his colleagues are aiming to produce a carefully calibrated WAIS that will enable this test to measure the factors in the Cattell structure more accurately than the simple compilation of the performance and verbal IQ.

The Stanford-Binet test

I shall now examine the items in the Stanford-Binet test (Terman and Merrill, 1960), the other individual intelligence test. Unlike the Wechsler scales the items are not formed into scales by type but are grouped together by level of difficulty. Only one score, the IQ, is thus obtained. It also differs from the Wechsler scales in that there is the one form suitable for ages of 4 to 17 years. This test is ideal for children but for adults of high ability there are insufficient items for good discrimination.

The types of items used in this test include: vocabulary, recognition of absurdities, being able to name the days of the week, abstract word definition, copying a chain of beads from memory, paper cutting, building a sentence from given words, digit repetition and the explanation of proverbs. This is only a sample of the many types of items comprising the Stanford-Binet.

There are several points to notice about these items. Some are virtually identical to those in the Wechsler scales – vocabulary, abstract word definition, digit repetition and the explanation of proverbs. The majority of the Stanford-Binet items are verbal, thus again favouring those from literate and educated backgrounds. With items like these it is almost certain that the test measures crystallised rather than fluid ability, although the few non-verbal types of items will tend more to measure fluid ability.

Since intelligence even in its narrow sense of the ability to educe correlates, its factor analytic definition, is a broad variable, the fact that there is a wide variety of items must add to the validity of the test. One final comment remains before I examine the factorial nature of the test. There must be some doubt as to whether scores of young children are equivalent to scores of the older children simply because the types of items used by the more extreme age groups are different. This is a problem with all tests of intelligence since young children obviously cannot attempt abstract definitions or complex logical problems. For example in the Stanford-Binet at the low age level we find items requiring the child to build a tower of blocks, identify parts of the body on a doll and name some simple objects. At the 12- and 13-year-old level subjects have to complete sentences, memorise abstract designs and make inferences to answer questions about a brief passage of English.

As should be clear from the description of the items, crystallised ability is the variable measured by the Stanford-Binet, if our analysis of the items and our definition of crystallised ability is at all correct. In fact, as Cattell and Johnson (1986) point out, this is exactly so. Correlations of between 0.76 and 0.82 have been obtained between the vocabulary scores and the mental ages derived from the whole test. Since, as I have argued above, vocabulary is the best measure of crystallised ability, there can be no doubt that this is what the Stanford-Binet measures. As such it is, indeed, an excellent test.

The British Ability Scales

These scales (Elliot, 1983) were developed to replace, in Great Britain, the two tests which I have just described. It was hoped that they would overcome many of the problems of the influence of social class and their theoretical limitations. However, the theoretical basis of the BAS was the work of Guilford and Piaget, work which militated against obtaining a clear measure of intelligence. Indeed it is recommended that the scales be given individually although customer demand has forced the test constructors to provide a brief set of scales from which a traditional IQ can be scored. I shall say little more about this test since it adds little to our understanding of factor analytic *g* (by design). This is because the tests selected to measure *g* are traditional high *g* loaders: speed of information processing, matrices (working out relationships among patterns), similarities, recall of digits, visual recognition, vocabulary and verbal comprehension. This is the usual mixture of crystallised (vocabulary and comprehension) and fluid ability (matrices) of traditional intelligence tests constructed before the modern factor analytic studies of intelligence.

These comments are not to denigrate the BAS, which is probably a useful measure of abilities which are valuable for educational psychologists to assess. It is simply that for the purposes of explicating the nature of *g*, for reasons that have been discussed, it is not enlightening.

Group intelligence tests

I shall now briefly describe and examine some of the best known group intelligence tests. Since there is a large number of these I shall restrict myself to those whose items explicate clearly the nature of fluid and crystallised ability.

Raven's Matrices

This test (Raven, 1965c) was developed just before the Second World War by Raven and has been subjected to considerable research and revision such that there are numerous versions, capable, in all, of testing the whole range of ability from 5-year-olds to superior adults. The matrices items are all of one kind, sequences of diagrams or patterns which require completion. What is needed to score correctly on the matrices is the ability to work out the relationship between the diagrams and to apply the rule to the new case.

These items are, without question, one of the best measures of fluid ability. Indeed almost all the variance in the matrices can be explained in terms of fluid ability and a factor specific to the item form, as can be clearly seen in Cattell (1971) or Cattell and Johnson (1986). These items can be constructed, relatively easily, with differing degrees of difficulty and it must be said that they can be extremely difficult, as any readers can attest who have attempted to complete the last items in the Advanced Matrices.

These items measure fluid ability because generally they present the problem in a form with which all subjects are equally unfamiliar, thus negating to some extent the advantages of social class and education, even though it helps somewhat to be verbal, naming the shapes and patterns makes it easier to work out the relationships. Thus although the matrices are useful in assessing the abilities of subjects from poor educational backgrounds, it has not been found that they are equally effective in cross-cultural testing, a topic which will be dealt with in a later chapter.

Although the matrices are excellent items, the Raven's Matrices on its own is limited simply because there is insufficient variety of items. Inevitably it will favour those high on the factor specific to the items and be biased against those who are low on this factor, which is, of course, irrelevant for measuring intelligence.

For this reason Raven advocated the use of vocabulary scales, as has been seen, the best measure of crystallised ability, and he has produced the Mill-Hill and the Crichton scales (Raven, 1965b, 1965a). Large differences in performance on the verbal and non-verbal tests can be revealing. It was also for this reason that Cattell in his Culture-Fair Intelligence Test used four other types of items, all equally unfamiliar or familiar.

The Culture-Fair Test

As with the Raven's Matrices the three forms of this test (Cattell and Cattell, 1959) allow subjects from a very young age (3 years) to superior adults to be tested. This test, specifically designed to be as culture-free as possible and thus to measure fluid ability, has five kinds of non-verbal items. All these, therefore, define the nature of fluid ability. The item types are described below, although the examples are not taken from this test, but are adapted from a previous study of intelligence test items (Kline, 1979).

Non-verbal items

Analogies These items are similar to verbal analogies but, being in non-verbal form, they are not dependent on a knowledge of language. The working out of analogies has been minutely studied by Sternberg (1977), who regarded their solution as typifying the processes of intelligent thinking. In general non-specialist terms (the work of Sternberg will be fully discussed in later chapters of this book) to solve analogies requires that subjects educe the relationship between the terms and apply it in the new case, a good example, therefore, of what Spearman meant by intelligence. It is interesting to note that classics demands highly similar abilities. Thus in translating from English, for example, the rule has to be educed, and then applied in the new case. What defeats the unintelligent is that each example is varied. Thus purpose can be expressed in many ways in English and the student who attempts to memorise the forms without comprehension is doomed. It is no surprise that ability in classics is highly *g* loaded. Figure 4.2 shows two examples of non-verbal analogies.

Series items In the non-verbal form these are similar to matrices. A series of patterns is shown and subjects have to choose the next one from a selection of possible responses. Clearly, to solve these problems, the rule underlying the series has to be worked out and applied.

Classification These items have the popular name of odd-man-out, because they require subjects to select from a list the one term that does not fit. Such items can be verbal or non-verbal, their solution being dependent upon the subject's ability to work out the classification that embraces all the terms but one. As with series, an example is not necessary.

Matrices These have been fully discussed in our study of Raven's Matrices and no more description is required.

Topological conditions This is a type of item which is mainly used by Cattell. Since these items are difficult to invent, my example is taken

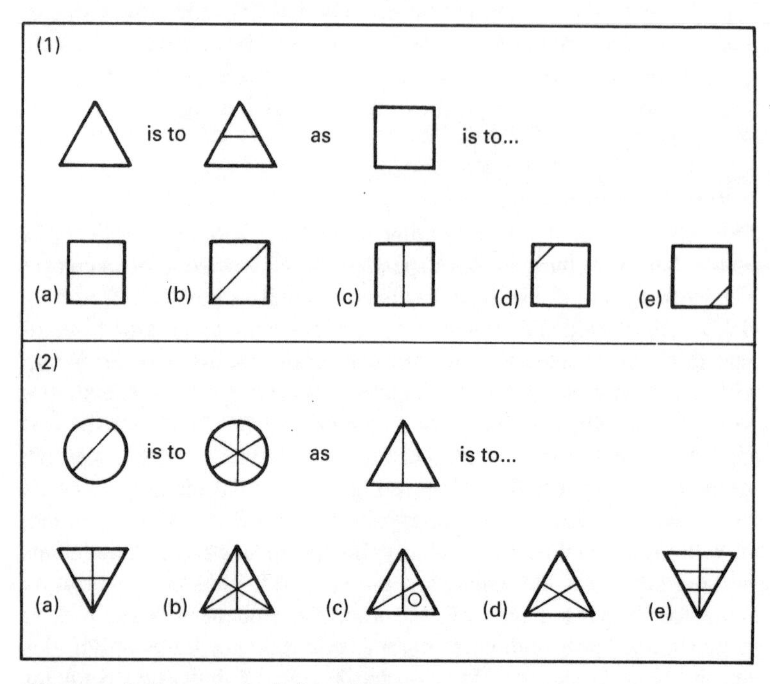

Figure 4.2 Two non-verbal analogies

from the Culture-Fair Test. Five boxes are drawn in which are circles and squares in various configurations. Subjects are required to indicate into which of these five boxes a dot could be placed so that it would be inside a circle but outside a square. As with all the items which I have discussed the solution depends on the eduction of relationships and application to new material.

Verbal intelligence tests

So far I have discussed two well-known intelligence tests that measure fluid ability rather than crystallised ability which for purposes of educational and vocational selection may be more useful. The highly able person who has not invested his abilities in the skills valued by a culture is not, almost by definition, so useful in a job and is unlikely to be successful academically. I shall now examine verbal intelligence tests or tests with considerable verbal sections. Most of the best measures of crystallised intelligence use the same kinds of items as those which have been discussed but in a verbal form.

Analogies

Verbal analogies are popular as measures of crystallised intelligence. One test, indeed, Miller's Analogies (1970), is used in the USA for selection for Graduate School in many of the best universities. There is one important point to note about verbal analogies. Their difficulty stems from two sources. One is the obscurity of the knowledge involved, the other the difficulty in working out the relationship between the terms. Two examples will clarify the point.

Samson Agonistes is to *Comus* as *The Bacchae* are to . . . and the choices are *Oedipus Rex*, *Medea*, *Cyclops* (correct answer), *Prometheus Bound*, *The Tempest*. To solve this item we have to see that the relationship of *Samson Agonistes* to *Comus* is one of common authorship and contrast. This done, the answer is obvious, given that we have the knowledge. Here then the reasoning is simple but the knowledge obscure. It is probably a bad item for measuring crystallised intelligence although it resembles many to be found in Miller's Analogies.

Television is to microscope as telephone is to . . . and the choices are amplifier, microprocessor, microdot, microphone, loudspeaker. This item is probably far superior as a test of crystallised ability, because its difficulty lies in working out the relationship between television and microscope, although it also requires some knowledge of modern electronic gadgets. Subjects have to see that the relationship is one of distance to size magnification in the visual field. Hence the correct response is amplifier. Notice that the distractors (incorrect responses) have been cunningly devised so that subjects who have worked out only a vague relationship would probably choose microphone.

As these two examples make clear, analogies are highly flexible items allowing for all kinds of subjects. It is easy to vary difficulty in terms of content and relationship. In addition to verbal analogies, analogies can be made with letters of the alphabet and numbers, all testing a combination of fluid and crystallised ability.

Classification items

These are popular as verbal items although they must demand knowledge to some extent in this form. However, they are more than tests of knowledge as two examples will demonstrate.

> Sparrow, starling, goose, bat, swallow. This is a simple example based upon the ability to classify into birds and mammals.

> Gull, swallow, swift, duck, starling. This is a more subtle item since the odd-man-out is now starling since all the rest are birds but have other meanings. If lark had been substituted for duck the item would have been ambiguous because gull is the only bird which swims.

Series

Verbal series items are possible although, in the main, they must test crystallised ability since much depends on a knowledge of language, as the following examples illustrate.

> minuscule, minute, tiny . . . big, large. The subject selects, to fill the gap, from: enormous, small, heavy, gigantic, prodigious.

These are the main item types used in crystallised ability tests, of which there is a large number. For full details readers should consult Buros *Mental Measurement Yearbooks* (e.g. 1978) to which I have referred. Tests which have good evidence for validity are: The AH series (Heim *et al.*, 1970), the Otis tests (1954) and the Moray House tests (Thomson, 1946), used for many years for selection for grammar schools in Great Britain, together with the group tests which have already been mentioned.

From this discussion of the best known intelligence tests and the items which comprise them I think that the nature of crystallised and fluid intelligence as it is conceived in factor analysis and psychometrics becomes clear. Essentially all these items are attempts to explicate, in operational terms, Spearman's original definition of intelligence. The fluid ability tests use, in general, non-verbal and unfamiliar materials, while crystallised ability tests measure intelligence embedded within intellectual tasks embodied in the culture.

One more point demands a brief discussion. Most intelligence tests have a factorial model underlying their construction. That is, it is assumed that the items load essentially on the intelligence factor which takes up most of the variance and that there may be some specific variance together with error. In good tests these last two sources of variance are relatively small. All the tests which I have described adopt this model with the exception of the British Ability Scales, some of which have a different basis. These tests are Rasch scales and these must be described.

Rasch scales

The Rasch (1960) model of responses to an item claims that the probability of a subject's responding correctly to an item is a function of two variables: the subject's status on the latent trait underlying the item, and the facility of the item for eliciting the trait. To take the relevant concrete example of an intelligence test item, then according to the Rasch model the higher the intelligence of the subject (the underlying trait) and the easier the item (high facility for eliciting the trait), the more likely it is that the subject will answer that item correctly. Put in this way, the model seems unremarkable.

However, what is remarkable about the Rasch model is that Rasch supplies methods of estimating subjects' status on the latent trait independently of the difficulty values of the set of items used. Similarly the difficulty value of the items can be estimated independently of the ability of the subjects doing the test. If these claims are accepted then it means that with Rasch scales exactly equivalent tests can be constructed, thus allowing accurate retesting and tracing of intellectual development. Furthermore these tests allow a true zero so to that extent they are independent of norms.

Given these almost magical characterisitics, and certainly ones that most psychometrists would desire, the question must arise as to why in almost thirty years there are so few published tests which utilise the technique. This is because, despite the staunch advocacy of Rasch scaling by Elliot in the British Ability Scales, there are various objections which have been made to the model, which are excellently summarised by Nunnally (1978).

Some of the assumptions of the Rasch model are almost certainly wrong. For example items are assumed to have the same discriminating power. In latent trait theory it is also assumed that only one factor accounts for item responses, an assumption that factor analytic studies contradict empirically, even of tests such as *g* which are predominantly unifactorial. Furthermore the influence of guessing is ignored in the model. In fact guessing can be accommodated but if this is done the model cannot be regarded as yielding item indices which are population free or subject indices which are item free, and much of the attraction of the model is lost.

There are yet other objections. Chopin (1976) points out that most Rasch tests have dealt with factorially simple variables where a simple latent trait may not be a gross distortion. However, in the construction of these tests it is difficult to obtain items that fit the model. In his work with the National Foundation of Educational Research in Windsor he even argued that if you test items sufficiently none fit the model and thus the formation of Rasch scales is impossible, since if sample free item statistics are impossible so are ability measures independent of items. This means that the only way Rasch scaling could be worked out is to calibrate the items on huge normative samples but this is little different from normal testing procedures, as is claimed by Lord (1974). Indeed the final objection (in my view decisive) against Rasch scaling is that if the items in an ordinary test are Rasch scaled there is a very high correlation between the two sets of scores, thus demonstrating that the complex mathematical procedures add little or nothing to validity (Nunnally, 1978). I think it can be concluded that for highly homogeneous tests, such as attainment tests in maths or chemistry, Rasch scaling might be a useful procedure for producing equivalent tests for a public examination. For psychological tests, however, of more complex variables, the use of Rasch scales is not supported by any evidence of increased validity.

Conclusions and summary

A study of the items in some of the best known and validated intelligence scales, both individual and group tests, has revealed clearly the psychometric, factor analytic concept of fluid and crystallised ability. The claim is, of course, that the abilities tapped by these items are common to almost all problem solving in the real non-test world which accounts for their ability to predict educational and occupational success. The nature of Rasch scales was examined but it was found that most of their purported advantages were illusory in that Rasch scales and normally constructed scales were little different.

Chapter five

Intelligence and educational and occupational success

In this chapter I want to counter, in part, the old objection to tests of intelligence, namely that they measure only skill at the tests. In factor analytic terms this means that the factor on which the items load is nothing more than a bloated specific, an impression which is reinforced to some extent by the definition of intelligence as what intelligence tests measure (Boring, 1923). Clearly if it can be shown that academic performance or job success can be predicted by tests of intelligence, it makes no sense to think of intelligence tests as measuring some specific factor.

Intelligence and educational success

There is a huge number of researches into the correlations between intelligence tests scores and academic performance, which is hardly surprising since the origins of intelligence testing lay in the effort to select children worthy of education, for example Binet and Simon (1905). However, as both Snow and Yalow (1982) and Jensen (1980) point out, this work is relatively easy to summarise simply because the results are clear, remarkably so by the standards of psychological research.

As Jensen argues, the results in support of a substantial correlation between intelligence test scores and educational achievement are so incontrovertible that critics of the tests accept them but interpret them not as evidence for a link between intelligence and educational success, but rather as a criticism of the tests themselves: that the tests reflect simple socio-economic advantages, or increased educational opportunities. Nevertheless it is the case that intelligence measured at the age of 5 predicts better than any other variable a child's future educational progress and attainment. To quote Jensen (1980, p.317) 'Children with higher IQs generally acquire more scholastic knowledge more quickly and easily, get better marks, like school better and stay in school longer.' At this point I shall draw special attention to the words quickly and easily, a notion of some importance in the ensuing discussion.

These claims require some further explication. The first point to note is that intelligence and educational success or achievement are not perfectly correlated. It would be surprising if they were since it is obvious that other variables than intelligence must play some part in how we do at school or college: how hard we work, how much we like the subjects, the quality of the teaching, our health and how we are getting on at home, just for example. Indeed given the influence of such factors as these what is surprising is that IQ scores should be correlated at all.

One argument that might be used against the interpretation of these findings as support for the importance of IQ in determining scholastic achievement is that IQ itself is also affected by all these variables, implying that IQ is no more than learning. However, examination of the content of IQ tests, described in the last chapter, shows that this interpretation will not hold, especially with tests of fluid ability where items are used which are quite unfamiliar to the subjects. There is no learning involved with Raven's Matrices.

It might be supposed that what subjects had learned was the ability to reason, and that it is this learning which is affected by all the variables which affect educational achievement and that this, therefore, accounts for the observed correlations. This argument, however, will not do. Achievement in education requires knowledge. It is this acquisition which is hindered by adverse environments. The capacity to reason is qualitatively different. The knowledge involved is small in quantity, and is far less likely to be hampered by environmental deficiencies than are normal educational objectives. Furthermore it is difficult to learn.

The evidence for this comes from several sources. First in an extensive review of research into the effects of coaching and practice in intelligence tests Vernon (1960) showed that effects were small: gains of up to 9 IQ points after about 6 hours' practice and that further practice brought little reward. Maximum effects are produced by explaining how items are done, showing the reasoning behind the correct responses. As is to be expected the effects are greatest with subjects who have little experience with tests and with non-verbal and performance tests. It is important to note that these effects do not generalise well. They are less with parallel forms of a test and even smaller with intelligence tests of a different type. All this suggests that it is difficult to improve the reasoning capacities involved in intelligence and that all that is being done is to increase the scores on the specific test variance rather than on *g*. This accounts for the lack of transfer to different tests.

These findings must be contrasted with studies of human learning where it is generally the case that performance improves with practice. Thus Ericsson (1988) presents considerable evidence that various memory tasks can be improved by practice, for example digit spans of twenty can be attained. However, there is little generalisability to other material. Similarly with meaningful material, if long enough is spent performance can be improved, as was demonstrated many years ago by McGeoch (1942), and everyone knows this from their own experience. This point will be further discussed later in this chapter.

If we put these sets of results together it is clearly incoherent to conceive of intelligence as attainment, although this is not to deny that highly adverse environments will affect performance on intelligence tests. The point is that such deficits are far smaller than with attainment tests and scholastic performance. Indeed as Vernon (1960) points out, one of the pioneers of the factorial analysis of abilities, Godfrey Thomson, developed the Northumberland Intelligence Tests for this precise reason. He found that in Northumberland mining villages children appeared backward on tests of attainment but bright ones could be spotted using intelligence tests. No doubt had these children had all the advantages of the upper classes, they would have scored even more highly but, as the research on coaching and practice indicates, the effects would be small.

From this I think it is fair to conclude that the substantial correlations between IQ scores and educational attainment cannot be explained away by arguing that intelligence itself is essentially an attainment. One further closely related point should be dismissed. This states that the content of the tests is highly similar to what has to be learned in education, and that this alone accounts for the correlations. In the case of the WISC and WAIS tests, as we have seen, there is some truth in this point. However, it should be noted that the offending scales, information and vocabulary, are more than counterbalanced by the other scales in the test which load also on the *g* factor. Certainly it is preferable to measure intelligence in a less culturally biased form and this is exactly what the fluid ability tests, developed factor analytically, do. It is here that my warning note about the notion of ease and speed of learning is important.

Vocabulary loads highly on *g* because the intelligent person learns vocabulary in normal everyday life faster than the unintelligent. This compensates, to a large extent, the greater exposure to words that a child in a literate home has, compared with a disadvantaged peer. Of course a very bright child in such an environment will have an advantage over a similar child from a poor environment. That is why, as was previously argued, if we have used a vocabulary test to measure intelligence then a non-verbal test should also be employed.

There is one other point. Acute readers may have noticed an apparent contradiction, namely that practice has little effect on intelligence tests and much effect on memory span. Yet memory span forms part of the Wechsler scales. However, digit span loads on the *g* factors on subjects who have not specially practised the ability. There is in unpractised subjects a correlation between the span and general ability. However, because the skill can be improved with practice it does not mean that it should not be used for intelligence testing. That it loads on *g* reflects again the fact that intelligent people learn more effortlessly and quickly than do the unintelligent. The only conclusion that can be drawn from this apparent contradiction is that for subjects who have been trained in digit span, digit span ceases to be a test of *g*. Practice has increased their digit span alone and has not rendered them more intelligent. Effectively, training acts upon the specific rather than the *g* variance in this test.

There is a further important point that bears upon the argument that intelligence tests predict educational performance because they measure some ability that is salient to such performance. This is the fact that IQ scores can predict performance in fields of which subjects are entirely ignorant and when there is no overlap of test item and educational criterion. It could be objected by those who were determined to deny, for personal or ideological reasons, that the notion of intelligence had any force, that such a correlation could be attributed to some motivational or personality factor, for example, trying hard. This argument is, however, entirely refuted by the work of Cattell and Butcher (1968). These writers demonstrated in a study of intelligence and creativity in which ability, personality and motivational factors were included that the personality and motivational factors were quite separate from ability factors, and that they were useful in predicting educational achievement because they were concerned with a different part of the variance from intelligence tests. Although, as I showed in Chapter 2 on factor analysis, there is a positive manifold among ability factors, that is they are correlated together, if the cause of these correlations lay in personality or motivational factors, the correlations would be greater than they are.

The median correlation between IQ and educational success is in the order of 0.5. In general it is highest among the youngest age groups and gradually decreases with age, a fact, again, which militates against the hypothesis that the correlation is due to some unknown personality or motivational factor. This makes good sense simply because, if we regard intelligence as of any consequence in learning, early in our lives it is of the utmost importance. Later such factors as how much we already know about a subject and how much we like it are clearly important and in this sense intelligence may be of less significance for adult learning than for what we learn in childhood.

The reasons for these changes, apart from the general point, discussed in the previous paragraph, are worthy of discussion, albeit briefly, because they throw light on the nature of intelligence as well as its influence on educational achievement. One possibility to explain these correlations is that the relationship between IQ and educational achievement is not linear. Thus there may be a point above which more IQ is not useful. This would effectively lower the correlations at the higher age levels. However, examination of the relationship between IQ and achievement at all levels of achievement shows this not to be the case. It is still an advantage to have an IQ of 160 compared with 140.

When considering the absolute levels of correlations between IQ and educational factors a number of simple but important psychometric points have to be borne in mind. The first concerns the reliability of the measures of educational achievement. The reliability of a measure limits the size of any correlation with it. Objective tests of educational success are more reliable than teachers' ratings and some intelligence tests are more reliable than others. What tests were used, therefore, must be taken into account when interpreting the results. Although statistical corrections for estimating the correlations with tests of perfect reliability are possible, I prefer not to use them, simply because they are only estimates.

Another, more severe problem lies in reduction of correlations due to homogeneity of variance in the variables, that is when the full range of scores is not found in the sample. This inevitably occurs at the tertiary level of education when only a relatively small proportion of an age group is involved. This factor alone would account for the reduction in size of the correlation coefficients in the older age groups. Although correcting factors can be applied once again it must be realised that these are only estimates.

It should be noted that IQ does not correlate equally with all subjects which can be studied at school and university. There are two reasons for this. In the first place some subjects require special abilities to a considerable extent. Music is the most obvious of these cases, although counterpoint and harmony, as Jensen (1980) argues, must be highly loaded on *g*. The complexity of the structures in Bach's *Art of Fugue* requires precisely the ability described by Spearman as the eduction of relations and correlates. The second reason is that some subjects can be mastered by prolonged practice and these tend to correlate positively but rather low with *g*. Indeed if we make a careful analysis of the abilities required for success in school subjects it is possible to predict those that will correlate highly with *g*. In general it is those subjects which require abstract reasoning and the ability to apply relevant information and make logical inferences. The traditional school subjects, natural sciences, mathematics, classics, where the level of difficulty depends on understanding what has gone before, fall into this category. History and the social sciences involve less of *g* than these subjects and this fact contributes to the lowered correlation between *g* and academic success at the tertiary level.

As has been argued, at the primary school the correlation between IQ and achievement is highest, around 0.6. Since progress at school depends more and more upon reading, it might be thought that perhaps this correlation was due to the fact that reading ability was helpful in scoring well on intelligence tests. However, performance in intelligence tests before children can read correlated highly with later reading ability, and with comprehension more than oral reading (Krebs, 1969). Indeed factor analyses of reading tests and IQ indicate that *g* is the best predictor of reading ability (Lohnes and Gray, 1972), even though there are some individuals whose 1Q is good but whose reading ability is extremely poor – the dyslexics.

If it is held that intelligence is a determinant of individual differences in educational success, as indeed the results reported in this section of the chapter are interpreted, then it might be expected that intelligence would be correlated with learning ability. Or perhaps it might be possible to re-define intelligence as speed of learning, in which case, of course, it becomes a redundant term. This point was briefly discussed previously but it warrants a little further examination.

Snow and Yalow (1982) have an excellent survey of this complex field which I shall briefly summarise. First all the research indicates that intelligence and learning ability cannot be simply equated. There is a complex relationship between them as is evinced by the fact that, as Cronbach and Snow (1977) demonstrated, estimated true mental age in any year is substantially correlated with gain in mental age in subsequent years. In other words, the child with high intelligence is likely to improve yet more relative to her less intelligent peers. Furthermore learning rate is a concept with little empirical support since any individual's rate differs for different tasks and even sub-tasks within a larger one, at least in the artificial, laboratory learning tasks favoured in experimental psychology.

This is essentially the conclusion of Jensen (1980) who argues that the rate of acquisition of conditioned responses, learning motor skills, simple discrimination learning and rote learning verbal material are little correlated with *g*, as is true for the retention of simple learning. If we remember Spearman's definition of *g*, this is hardly surprising. Furthermore if we consider what most people actually learn in the course of their education and even more their lives beyond the confines of institutions of education, then it is clear that these learning tasks are hardly relevant to real-life performance.

Jensen (1980) argues from a survey of the research on learning and its relation to IQ scores that some generalisations can be made. Thus learning does correlate with IQ when it requires conscious mental effort and is intentional, compared, for example, with conditioning processes which go on without awareness. As was the case with school subjects, learning hierarchically organised material, where later learning is dependent on what has been learned before, is correlated with IQ. Similarly learning meaningful material correlates with IQ scores (as distinct from the lists of nonsense syllables found in laboratory experiments) presumably because the intelligent person can make more connections with what he already knows and he already knows more. Transfer of learning from one task to another similar one is also correlated with IQ. It should be pointed out, in reference to this point, that almost all real-life learning is of this kind in that there is almost nothing which we have to learn that does not resemble something else, at least in part. It could also be argued that transfer could be facilitated among intelligent individuals because they can see similarities which the dimmer people fail to observe.

Further factors which tend to produce correlations between IQ and learning can be listed. The more that insight is required for learning the more IQ is involved. The complexity and difficulty of the task is important. Here moderate levels demand intelligence. If the task is too complex and difficult subjects regress to more primitive methods of problem solution. Learning and IQ are correlated when there is a fixed time for learning. When age affects the ease of learning material, learning and IQ are correlated. For example learning the names of favourite sportsmen is not age related and correlates almost zero with IQ. However, this is not true of reading or mathematics. We saw this phenomenon exemplified in the *g* loadings of school subjects.

Jensen (1980) makes a further most important point, namely that learning and IQ are correlated more at the initial stages of learning something new than at later stages, where gains are the results of practice. This is best exemplified by music. Learning to read music is positively correlated to IQ. However, individual differences among skilled practitioners, for whom sight reading is automatic, are not and this is true of many similar skills such as typing or morse.

The educational implications of these points are considerable. If we consider the conditions in which correlations between IQ and learning are found, it can be seen that these arise in most educational institutions. In learning school subjects, under normal circumstances, intelligent children, as defined by tests of fluid and crystallised ability, will be quicker learners and dim children will be struggling and will fail to master new subjects. This is why teachers find the notion of intelligence so appealing. Unlike the experimental psychologists they see the influence of intelligence everywhere in their jobs and the psychometric concept fits their experience closely.

Some of the objections to intelligence tests have concerned their use which has been branded as cruel and somewhat fascistic, as, for example, by Kamin (1974) in rejecting immigrants in the USA and by Pedley (1953) in selection for grammar schools. However, to condemn intelligence tests because they have been misused or because one does not like their use is absurd, no different indeed from condemning electric drills because they are used to injure kneecaps by the IRA. The point here is that intelligence tests are not necessarily bad, unlike bombs or bacteria developed for offensive purposes.

All this presupposes that there are good uses for intelligence tests in education and these must now be briefly discussed. First it has to be said that the objection raised by Pedley (1953) to the use of intelligence tests for selection for grammar schools was powerful, namely that the effect on those who were not selected was devastating. To be labelled a failure at 11 or indeed at any age is not good. I accept this point, even though, as a psychometrist, I am an advocate of intelligence tests.

Nevertheless intelligence tests have some important functions in education. In the first place if children are failing at something, then it makes sense to test their intelligence. As all the research discussed in this chapter shows, failure or poor progress may be the result of low intelligence. I shall illustrate the argument with a few examples, some based upon my own experience in Aberdeen. In that city there was great pressure placed upon children, especially from the educated middle class, to study hard and win a place at one of the fee-paying schools from which many eminent scholars have emerged. Many of these children, pushed beyond their limits, turned up at the doors of the educational psychologists with all kinds of problems. In many cases it was clear that too much was expected from them. When the pressure was eased, improvements were noticed. Just as labelling children as failures is cruel and profoundly unhelpful so too is driving children beyond their capabilities. Similarly the doctrine that all differences in performance depend not upon intelligence but opportunity and effort is bad for parents. If their children are not stars they feel guilty for what they have failed to do, and as I have argued the child, trying all he can but not succeeding, feels bad as well.

In other words if we know what to expect we can tailor a proper educational programme for a child. This is surely a sensible policy, better for children, teachers and parents. Used in this way to investigate failure, rather than as a barrier to opportunity, intelligence tests can be humane and useful instruments. There is one final point which has been mentioned in this chapter, but which requires some emphasis. Although it is impossible to eliminate all the advantages brought by a good home and a literate background in the development of a child's abilities, the assessment of educational potential by intelligence tests is certainly the least biased procedure. Thus to abandon the use of intelligence tests in assessments is a denial of elementary social justice for the poorer sections of the community. Far from working against the interests of the disadvantaged, the intelligence test is their best hope of gaining any recognition for their abilities.

Conclusions

All these arguments support a number of simple points.

1. It is not true that intelligence tests measure only ability at intelligence tests. If this were so there would be zero correlations with academic success.

2. The correlations between intelligence tests and educational success cannot be explained away as reflecting some common personality or motivational factor which affects both scores. The spheres of ability, personality and motivation are largely separate.
3. These correlations cannot be explained by some common content shared between educational criteria and intelligence tests.
4. Intelligence tests do not simply measure the ability to learn or speed of learning. The lack of correlation between rote learning of meaningless materials and IQ demonstrates this and is further support for point 3.
5. In fact IQ tests measure the ability to reason, to educe correlates, what Spearman, indeed, defined as intelligence. It is this ability which accounts for the correlations between intelligence test scores and educational achievement, correlations being highest where subjects demand reasoning and where there is little emphasis on learning. IQ tests are the least biased assessment of educational potential.
6. The ability measured by intelligence tests is essential to the mastery of new problems. That is why the correlations with subjects requiring a huge commitment and learning, such as music, are not higher. This aspect of learning is a function of motivational and personality variables.
7. The importance of intelligence in educational achievement is the emphasis of this chapter. However, as Chapter 3 on the structure of abilities shows, some special abilities are also important, music being an obvious case.
8. That IQ tests measure an ability essential for the solution of new problems is the reason why these tests are useful in occupational selection, where the skills required are unknown to the candidates such that attainment is not a good guide. It is this research that will now be examined in the second half of this chapter.

Intelligence and occupational success

Just as intelligence tests have been widely used by educational psychologists because they correlate well with educational achievement, so, also, they have been employed by industrial and occupational psychologists because they are correlated with occupational success. They play a part, therefore, in selecton. In addition to this they can be used in vocational guidance (see Kline, 1975). However, before the findings are scrutinised a few more general points need to be made.

A major difficulty concerns the measurement of occupational success. As Ghiselli (1966) has argued, in a book to which this section of the chapter owes much, comparability between studies is often dubious because the criteria of job success are different. The most commonly used criterion is the supervisor's rating. However, as a form of measurement this is poor. Ratings are notoriously unreliable (Vernon, 1961), reflecting the biases and prejudices of the rater as much as the performance of the subject, unless they are carried out by highly trained practitioners which, for obvious reasons, is not the case with supervisors. In addition, there are technical psychometric difficulties with rating scales which render their use dubious. Just for example, some raters use the extremes while others never do so. This alone can distort results. The only positive feature of these ratings is that, in most instances, the supervisors really know their subjects well.

Sometimes, however, in place of ratings and depending on the particular job, more objective indices are used. These include output, errors, wastage, sales volume and number of new accounts. These are superior to ratings but comparability between jobs is difficult and some occupations do not lend themselves to such objective criteria. Let me take two examples, to illustrate the problems. If we were judging car salesmen, then sales volume might appear to be an excellent criterion. However, this itself is influenced by several factors which render it a far from perfect criterion. For example we can use it as a criterion only for salesmen selling a particular make of car. A Ford salesman is bound to sell more than a purveyor of Bristols. If we changed to a financial criterion, the error would be in the opposite direction. However, even if we restrict the study to salesmen of a particular make of car, sales volume may not be a good criterion. If we take a relatively expensive car the location of the garage is bound to affect sales. Hence those in a good area will have an easier task and even the location of the garage of a competitor and the skill of their sales staff can affect results. These points have been raised simply to demonstrate that even an apparently good criterion of occupational success, such as sales volume, is not quite what it appears.

The second difficulty concerns the fact that many jobs cannot, from their nature, be assessed objectively. Teaching is a good example. Exam results are not a fair criterion because much depends upon the facilities of the school and the characteristics of the pupils. Furthermore there is more to an education than passing examinations. If we think what we would like our teachers to do we can see, immediately, the general cause of the difficulty with a simple objective criterion. The more multi-dimensional the requirements of a job, the harder it is to develop any kind of objective criterion and, of course, the more difficult it is to rate the individuals in that job. The only way around these problems in the study of job success is to use as many and as diverse criteria as possible.

One other problem must be mentioned. In some researches the data were obtained after or even during training. This is particularly so with ratings or test scores. Now, as we saw with music, it can be the case that acquisition of a skill requires characteristics different from those necessary for its maintenance. There are other difficulties that must not be ignored in establishing criteria for the purposes of either selection or guidance. These concern the fact that the psychological characteristics of jobs may change as individuals are promoted. Thus, as Dixon (1976) demonstrated, the qualities demanded for a good young officer are actually antithetical to those required by a successful general.

All these arguments are not to write off the work in which correlations are presented between intelligence and occupational success. On the contrary they are intended to enable due caution to be shown in interpreting the findings.

The findings

As I indicated earlier in this chapter, this section of the chapter is indebted to the work of Ghiselli (1966) as, indeed, must be all who attempt to write on the relationship between intelligence and occupational success. This is simply because in his book Ghiselli surveyed and pulled together the results of 10,000 investigations relating human ability to job success. Recent work by Gottfredson (1986) simply confirms all these points.

One fact stands out, a finding of special relevance to this chapter. Ghiselli (1966) showed that *g*, as measured by a variety of intelligence tests, correlates on average 0.3 with success in any job. In other words if one were forced to select people for unknown jobs, an intelligence test would allow one to do so significantly better than chance. No other ability variable achieved an average correlation of this size.

From the viewpoint of selection and guidance these results show that for a large number of jobs intelligence is a good predictor and thus intelligence test scores would be essential information. However, from the viewpoint of this chapter, where I am trying to establish that the factors measured by intelligence tests are not specific to the tests but are, as factor analysts have argued, general, these results are highly salient. They demonstrate beyond all equivocation that fluid and crystallised ability are not test specific but general across all occupations. Thus the argument that high scores on IQ tests measure nothing but test ability is wrong, utterly and completely.

The highest correlations between job success and intelligence were between 0.6 and 0.7 and one interesting point emerges from these correlations, compared with the correlations with narrower aptitudes or group factors. Group factors are important. For example where verbal ability is of prime importance, as in journalism or lecturing, it predicts better than intelligence. This is hardly surprising. However, it does not mean that aptitude tests, measuring group factors, are more useful for selection or guidance than intelligence tests, for a number of reasons that will be discussed below.

First it is necessary to define aptitudes. As I have argued in considerable detail in previous publications (Kline, 1975, 1979) the term aptitude has two distinct meanings, a fact which causes considerable confusion. In some instances aptitude refers to a clear group factor of the kind discussed in Chapter 3. Verbal or numerical aptitude would be examples of these. However, in some cases aptitudes are far from factor pure and are of little psychological interest, being nothing but sets of abilities and, sometimes, even personality traits, which happen to be important for some occupations. An example of this type of aptitude is clerical aptitude, which is a mixture of speed and accuracy in computation and copying, conscientiousness, vigilance, resistance to boredom and a love of orderliness for its own sake.

The main reason that aptitude tests are not highly valuable in selection procedures is due to the fact that much of their variance is accounted for by the general ability factors, fluid and crystallised ability (Quereshi, 1972). This means, for example, to take clerical aptitude again, that the highest scorers on the clerical aptitude test are not clerks but individuals with high *g* and conversely clerks do not have their highest scores on the clerical aptitude test. In brief the same individuals tend to get the highest scores on all the factors, thus differential aptitude tests are not really differential at all but measure *g*, somewhat inefficiently. Far better to measure *g* separately and then measure any relevant group factors.

The only way in which aptitude tests could be used for selection is to carry out large-scale studies of large numbers of occupations and draw up the best weightings of variables for predicting success in the different occupations, using multiple regression. If this is done then aptitude tests can add in useful variance over and above the measures of *g*.

However, these points are peripheral to the main thrust of this section of the chapter, namely that studies of occupational success have shown that the general abilities measured by intelligence tests are related to occupational success, however it is assessed. All this is evidence that intelligence tests measure factors important in real life and not restricted to the tests themselves. Thus Boring's (1923) definition of intelligence, as what intelligence tests measure, can be seen in the light of this evidence, concerning occupational and academic success, not to be as absurd as it first appears.

So far I have discussed the relation of intelligence to occupational success, but there is other evidence relevant to the ubiquity of intelligence in real life, in the occupational sphere. Thus, as Jensen (1980) points out, there is a substantial correlation between the prestige of an occupation, however it is rated, and the IQ of individuals in the occupation. It is curious that, if IQ scores are nothing but ability at IQ tests, ratings of jobs should correlate with such an objective index.

One possible counter to this argument is that these 1Q differences between occupations are really educational differences, since it is the case that the occupations with the high IQs demand individuals with considerable education and the converse is true at the other end of the scale. This argument, however, will not stand. A huge study of 10,000 air force cadets by Thorndike and Hagen (1959), who were tested at 21 years of age and whose occupations were noted 12 years later, indicated clearly that those in better jobs had higher IQs than those in lower ones. Furthermore this sample was restricted in range in that all of them were of above average ability. There can be no doubt that high IQ is important in attaining high-prestige, professional jobs.

Some other points are worthy of note. For some jobs a high IQ is a necessary but not sufficient condition. It is simply impossible to be a theoretical physicist without a high IQ. This means, in fact, that low IQ is more predictive of occupation than is high IQ. High IQ individuals can be found in all jobs whereas in occupations of high status this is not so.

Of course this is not to deny that social and hence educational differences affect the jobs that people have. This is manifestly the case especially in Great Britain, where attendance alone at one of the great public schools is almost a *sine qua non* for some occupations. Certainly high IQ children from families of low socio-economic status find it more difficult to attain high-prestige jobs than children of similar ability from more privileged homes. Yet, as the work of Terman and Oden (1959) who followed up children of high IQ (mean 152) showed, in this gifted sample length of education and income were not closely related: those who had not proceeded to college were not inferior to those with bachelor's degrees. However, caution has to be shown in generalising the results from so exceptional a sample to the general population. Nevertheless it should be pointed out that the children in this group were more successful in their careers than groups comparable in social class and education but differing in IQ.

There is one further finding which finally destroys the claim that holding a prestigious job is determined by social class and all that goes with it, rather than IQ. Jenks (1972) showed that there is a low correlation between the occupational status of brothers of about 0.3 while Waller (1971) found that differences in the occupational status of fathers and sons were correlated 0.37 with intelligence. Clearly IQ and social background are determinants of one's occupation.

Conclusions

The conclusions from the arguments in this chapter are clear. Fluid and crystallised ability, as measured by intelligence tests, play an important part in academic and occupational success. This fact demonstrates that these abilities are, indeed, general and are not specific to the tests themselves.

Chapter six

Heritability of intelligence

The heritability of intelligence is a subject which, unlike many academic topics, creates deep feelings in those who become engaged in it. To introduce the topic I need to make a few general points and to reduce the emotional aspects of the argument to ensure a dispassionate view.

I intend to examine the evidence from one particular approach to the subject, that of biometrics. This is the study of the sources of variance within populations with reference to genetic and environmental components. It must be noted at the outset that when the findings indicate, for example, that 50 per cent of the variance for a given trait, perhaps intelligence, is attributable to genetic factors, it does not mean that half of the intelligence in an individual is genetically determined and the rest by her environment. These figures refer to the sources of variation within the population. Similarly if within a population some trait is genetically determined it does not mean that differences between populations on the trait must be so determined, as Plomin (1986) has explicated. As will be seen later in this chapter and in other sections of this book, these points are important.

This topic appears to be factual, in that it is an empirical matter, partitioning the variance into genetic and environmental components. Yet there are huge differences in conclusions ranging from Kamin's claim (1974) that there is no evidence that a rational person could cite to indicate that genetic factors determine individual differences in intelligence to the work of Cattell (1971) who would argue that about 70 per cent of the population variance is genetically determined. So that the argument, for the reader, does not become one simply of belief it will be necessary to set out the bases of biometric methods, the logic of which is remarkably clear.

However, there are not simply differences of opinion about the results of these studies of the determinants of the variance in intelligence. Writers disagree concerning their value. For example Anastasi (1961) writes that the nature/nurture question is of little importance and that what one wants to know is not the extent to which environment affects intelligence but how it does so. This argument is, of course, wrong in principle. If intelligence were entirely genetically determined the question which Anastasi believes to be important becomes meaningless. It is important only provided that we know that there is some environmental effect, of some useful size. The view of Anastasi is also echoed by Scarr and Carter-Saltzman (1982) who regard the field as of interest simply from the viewpoint of intervention.

My interest in this topic is not only practical (obviously interventional procedures are of great interest and social significance) but also theoretical. Our notion of intelligence as a variable must change if we were to discover that it is largely heritable or mainly determined by life experience. Furthermore the facts in this field are of crucial significance to the thesis of this book, namely that the variable measured by intelligence tests is of considerable psychological significance and is not a factor specific to the tests. If it emerges that the variance in intelligence tests is considerably determined by genetic components then the argument that it is test specific is rendered weak. It would be a curious matter if an ability that was restricted to a particular set of items and was of no psychological significance were largely heritable.

There is one further point which must be mentioned briefly at this juncture so that it may be dismissed from the arguments in the rest of the chapter. This concerns the alleged fraud of Sir Cyril Burt, who, it is claimed, essentially invented a set of data which demonstrated from a sample of identical twins reared in separate environments that there was a large genetic component in the determination of individual differences in intelligence. The evidence has been carefully sifted by Hearnshaw (1979) whose case appears overwhelming although recently the debate has been reopened (Fletcher, 1987; Kline, 1987). It is a pity that the accusations were not made before Burt died so that he had a chance to answer. None of Burt's evidence will be used in this chapter. In any case it would not be necessary since there are plenty of studies showing the same findings.

Biometric methods

First I shall set out a simplified algebraic explanation of these methods. This is based upon the brilliant exposition by Fulker (1979), of which the clarity is difficult to better.

> $P = G + E$ where P is the phenotypic variance, G, the variance determined genetically and E, the variance environmentally determined. By using variances in the model it is possible to separate out G and E from the variances and covariances of groups of individuals, such as twins.
>
> The analysis of variance of twin pairs partitions the variance into two sources: between and within pairs. The more pairs resemble each other, the greater the between pairs variance will be compared with the within pairs variance. In fact the ratio of $(B-W)/(B+W)$ yields the intra-class correlation showing how similar pairs of twins are. From these variances and correlations, the genetic and environmental components can be derived.
>
> E, the environmental component, can be broken down into two parts: the common or shared environment (CE) reflecting the experiences of home life that are common to members of a family, and the specific environment (SE) or unshared aspects of experience. Then the equation becomes: $P = G + CE + SE$. With this model the following arguments can be made.
>
> The correlation reflects the variance of all shared influences. For identical twins, $r = G + CE$; for non-identical twins, sharing half their genes, $r = \frac{1}{2}G + CE$. From these assumptions the following estimates can be made: G = twice the difference between the two correlations; CE = the difference between the MZ correlation and our estimate of G; $SE = 100 - G + SE$.

This is the basic reasoning behind the biometric approach to the analysis of genetic and environmental determinants of variance in a population for any trait. This is the simplest additive model and complexity may be added by using the intra-class correlations between other relatives than twins, and allowing, in the model, for dominance and assortative mating, for example. These, however, are complications beyond the scope of this chapter, where my intention is for readers to be able to see how it is possible to estimate hereditary and environmental components of population variance.

As can be seen, biometric methods determine the sources of variation within populations by examining variance and covariance between individuals of all degrees of relatedness, reared together and apart. As Jinks and Fulker (1970) point out in their lucid discussion of the biometric approach, biometric methods were developed for agricultural and biological use, although there is no reason to suppose that they are inadequate for the study of human beings. It is clear, from our description of the model, that twin studies form one aspect of biometrics but on their own they are inadequate as sources of knowledge for a number of reasons. In the first place studies of identical, monozygotic twins, reared apart, are likely to underestimate the contribution of genetic factors. This is because there are sources of discordance among identical twins which are neither environmental nor genetic in the ordinary sense of the words but which would be classified as environmental in twin studies where all difference must be so attributed. These special factors, which have been discussed by Darlington (1970), include nuclear differences which arise by gene mutation or by chromosomal loss or gain when the zygote splits; cytoplasmic differences which are brought about by deleterious genes acting differentially on the two organisms; embryological differences created by errors arising from a late splitting; and, perhaps the most common, nutritional differences due to unequal placentation.

There are other difficulties which render the results from twin studies dubious. For example the number of identical twins reared apart is obviously small. It is pertinent to consider whether these are a good sample of identical twins so that generalisations to normally reared twins could be made. Even more significant is the problem of whether it is safe to generalise from twins to singletons. Finally the twin method assumes that the genetic and environmental influences are uncorrelated and that they combine in a simple additive way without interaction. On intuitive grounds this assumption is dubious because it is likely that the environment of intelligent individuals is different from that of the less intelligent. For example intelligent children respond differently from the less intelligent and thus provoke different responses from those around them.

In biometric methods, as was demonstrated, the total variance is broken down into the between families and within families variance. In addition the contribution of the interaction of the genetic and environmental variance can be taken into account, as well as correlations between these two sources of variance. Indeed as Jinks and Fulker (1970) argue, one of the great advantages of biometric methods is that various genetic models can be tested. For example one can see whether there are effects of correlated genetic and environmental factors, of interactions and of dominant and assortative mating. There is no need to assume a simple additive, linear model.

I have always argued (e.g. Kline, 1979) that the results of these biometric analyses, where clear models of the effects of genetic and environmental factors are put to the test, are difficult to impugn, especially since the approach has been shown to be valuable in agriculture and horticulture. However, Feldman and Lewontin (1975) have argued forcefully that analysis of variance cannot really separate variation that results from environmental fluctuation from that due to genetic segregation. Even so, their arguments do not apply to biometric methods of the kind discussed by Jinks and Fulker (1970). Thus Feldman and Lewontin (1975) claim that broad heritability, total genetic variance, is not a useful statistic in human population genetics. What is required, they argue, is the narrow heritability, the proportion of variance due to additive genetic variance. In the biometric methods discussed in this chapter, both broad and narrow heritability may be computed. Indeed, Jinks and Fulker (1970) actually point out improvements in the calculation of broad heritability.

There still seems no reason to ignore the biometric work on intelligence which, if it were a variable of no psychological or physiological significance, would simply not be heritable. Before considering the findings one further point deserves a brief mention. There can be no doubt that measuring environments is an art or science even less developed than measuring psychological traits. This is partly due to the fact, of course, that the environment is not an objective but a subjective phenomenon. This has led some critics to the mistaken notion that estimates of the environmental component of the population variance must be flawed. This is not so since they are derived, in essence, by including in them what cannot be attributed to genetic factors, as was made clear by the introductory algebra. Thus no environmental measurements are necessary.

Results of the biometric study of intelligence

In a chapter of this length it is impossible to mention let alone critically discuss all the results of biometric research, using the term in the broadest sense to include twin studies. Scarr and Carter-Saltzman (1982), for example, take almost 100 pages and cite more than 200 references and that was more than 10 years ago. In this chapter, therefore, I shall attempt to summarise the main thrust of the findings and examine any anomalies or differences in interpretations of the studies.

The first point to emphasise is one that is usually made in discussions of this sort but also usually ignored. This concerns the fact, as Plomin (1988) argues, that all the findings, especially the most recent, indicate that there is strong evidence for environmental as well as genetic influences on the development of intelligence. In addition there are some highly interesting and important findings concerning the way in which the environmental factors operate.

I have already made the point that the relevance of the question concerning the genetic and environmental determination of intelligence, and indeed any psychological trait at all, is not just practical but theoretical. In fact Plomin (1988) spells out exactly some of these issues which are implicated in the biometric approach to the study of individual differences. Some of these points have been discussed and include: genetic differences among individuals can lead to phenotypic (i.e. expressed) differences. To the extent that heredity is important, the expected phenotypic similarity among relatives is a function of their genetic similarity. Environmental differences among individuals can lead to phenotypic differences. The environmental component of variance can be decomposed into two; one shared, the other unshared by family members. Quantitative genetic parameters will change when genetic and environmental sources of variance change. Heredity can mediate relationships between environmental and behavioural measures. In addition to the main effects of environment and heredity, phenotypic variance may arise from an interaction of genotype and environment. Similarly such variance can arise from the correlation of genotype and environment. Correlations among traits can be mediated both genetically and environmentally and both environmental and genetic factors can be responsible for continuity and change.

It is to be noticed that some of these assumptions are obvious and are barely worthy of note while others are of more psychological interest. For example the splitting of the environmental determinants into shared and non-shared components is highly significant and its implications will be examined later in this chapter. Similarly the fact that if sources of variance change, parameters change, means that extrapolations from one population or culture to another are unsafe. In India, for example, where there is a far greater variance among environments than there is in Great Britain the relative importance of the environmental component in the population variance must be greater, assuming that there is similar variance in the genetic component.

As Plomin (1988) points out, if we consider twin studies, family studies and adoption studies, all aspects of the biometric approach, there are clearly different problems with each of them. The difficulties with twins have been mentioned but the problem with adopted children again concerns the difficulty of extrapolation to natural children and the fact that it is possible that adopted children's relationship with their adopted parents is different from that of natural children as, too, may be their relationship with adopted siblings. Given all these factors, that there is any agreement at all between findings is remarkable and gives one confidence in the results.

Plomin (1988) presents a table comparing old and new studies of the correlation between individuals of different degrees of relatedness reared together and apart. The new findings are highly similar to the old but show slightly less heritability. This, however, is certainly due to the test used, which was an attainment test, the National Merit Scholarship Qualifying test. Its use as an IQ test is defended by Plomin on the grounds that it contains a vocabulary test. However, as we have seen from Chapter 3 on the factor analysis of abilities and Chapter 5 on the relation of IQ to educational success, it is obvious that this test is not a good measure of IQ, and I have more confidence in the earlier figures, which are those of Erlenmeyer-Kimmling (1963) with the Burt data removed.

Table 6.1 shows the correlations between relatives for IQ. These correlations (in the old studies) indicate a high degree of heritability for IQ, the identical twins reared apart and the siblings reared apart being particularly significant. These correlations indicate a heritability for IQ in the population of approximately 70 per cent. When it is realised that siblings share only half their genes, the 0.4 correlation is remarkable. Plomin (1988) tries to argue from the newer figures and from more recent studies, not included in the table, that this heritability estimate is too large, around 50 per cent being a more accurate figure. However, as I have suggested, the newer figures really refer to attainment not intelligence and the newer studies were based on very small samples. Moreover, even if we take the conservative figure of 50 per cent it means that by far the best predictor of IQ is the biological parents' IQ, even of adopted children. As Plomin (1988) points out, IQ can be thus predicted with a standard error of only 12 points. This estimate could be improved if we accept the higher figure of the true IQ studies. Nevertheless what these figures also demonstrate is that the environment too plays an important part. Indeed if we take the lower estimate it ought to be possible to predict as accurately from measures of the environment. However, these influences are so diverse and complex that this is far from the case.

Table 6.1 Twin data: correlation coefficients for old and new IQ data

	Old data		*New data*	
		Number of		*Number of*
	r	*pairs*	*r*	*pairs*
Genetically identical				
Same individual tested twice	n.a.	n.a.	0.87	456
Identical twins reared together	0.87	1,082	0.86	1,300
Identical twins reared apart	0.74	69	n.a.	n.a.
Genetically related (first-degree)				
Fraternal twins reared together:				
Same sex	0.53	2,052	0.62	864
Opposite sex	0.53	(total)	0.62	358
Non-twin siblings reared together	0.49	8,228	0.34	776
Non-twin siblings reared apart	0.40	125	n.a.	n.a.
Parent–child living together	0.50	371	0.35	3,973
Parent–child adopted apart	0.45	63	0.31	345
Genetically unrelated				
Unrelated children together	0.23	195	0.25	601
Adoptive parent/adoptive child	0.20	n.a.	0.15	1,594
Unrelated persons reared apart	–0.01	15,086	n.a.	n.a.

Source: (Plomin, 1988)

What needs to be stressed in consideration of these heritability figures is that even the highest of them (70 per cent) does not rule out considerable influence of the environment. As Jensen (1989) demonstrates in a review of the Milwaukee project to improve the IQ of children at risk in low IQ families, work that I shall examine later in this chapter, with 70 per cent heritability, environmental differences can produce up to 38 points difference in IQ. Nevertheless it can be concluded from a study of these correlations between relatives living together and apart that there is a considerable proportion of the population variance in IQ determined by genetic factors, at least 50 per cent and probably more in the region of 70 per cent. Parents' average IQ is the best predictor of a child's IQ regardless of where she is brought up. However, since these figures indicate that there are important environmental influences on intelligence I shall now examine what these are and how they work.

Environmental variance

As has been shown, the contribution of environmemtal factors to population variance is by subtraction, a simple additive model being used, where the total variance consists of genetic plus environmental variance. However, as was also demonstrated, the environmental variance is broken down into two components, the shared and the non-shared environmental variance, and these terms must now be scrutinised with great care. Included in the shared environment, defined as what makes family members similar to one another, are such variables as parental income and education and those aspects of school and family which are similar for all the children in the family. The non-shared environment is that aspect of experience which tends to make children dissimilar: different reactions to different children, differences due to birth order exemplify this group of variables. The problem here with these definitions is that it is difficult to say, a priori, what would make children similar to each other. It could be argued by those who stress the subjectivity of experience that nothing is common to any individuals, even identical twins who do usually have highly similar upbringing. Nevertheless, despite these difficulties the distinction is useful. Certainly research suggests that it is.

With reference to investigating the effect of the environment, adopted children are ideal since any resemblance must reflect the shared environment, by definition. Studies of the correlation of IQ between unrelated children reared together show a correlation of around 0.3 to 0.35 if small sample studies are ignored. However, by adulthood these correlations have dropped to zero. Unrelated adults who have been reared together are no more alike in intelligence than are random pairs of individuals. This is a truly contra-intuitive result, as Scarr and Weinberg (1978) admit. The longer adopted children stay in the same family the less alike they get.

Plomin (1988) argues that these data are consistent with the claim that environmental factors contribute to about half the variance in the population IQ but that in the case of adults it is the non-shared environment. It is to be noted that this explanation is tenable only in the light of the fact that the correlation between identical twins reared together is not 1. If it were, a more simple explanation of these findings is that family influences have no effect on intelligence. To claim that the family influences that affect this correlation are those that make people dissimilar is necessarily true, from the meaning of the terms, and is a non-contingent statement of little scientific interest (Smedslund, 1978).

If we accept Plomin's explanation of these findings then it has to be said that it is the non-shared environment within the family, not the shared environment, that influences the development of intelligence. Thus common family influences such as the education of the parents, how much they discuss stimulating topics in the home, how many books there are about, do not seem to matter. What is influential is the particular interaction of the parent with each child, an interaction which is necessarily unique even for identical twins, a uniqueness brought about by the way a child responds, by his relationship with the other parent and other children, if present, and by how the parent feels towards the child. What is unclear about these findings, however, is why it is that these unshared aspects of family life should be of overriding importance. It is also unclear as to what these unshared environmental influences consist of, the factors which I have mentioned being no more than likely examples.

Another possible and obvious explanation, and one which accounts for the fact that the correlation between adopted and natural children decreases with age, is that as the children grow older the influence of the family becomes less and the effects of school and peers takes over. If an adopted child is very different from the other children in the family he or she will have different friends and do different things and such difference will be increased by these non-shared aspects of the environment. This is certainly what happened in the case of my own adopted child who, almost deliberately, created for herself a lifestyle which was the opposite of that of the rest of the family.

The psychological and practical importance of the finding that the non-shared environmental variance is influential on the development of intelligence is considerable. If it is true, family intervention projects, for example, are doomed to failure. So too are school projects, for these would constitute shared environments for members of a family. It is therefore essential to examine the meaning of shared and non-shared environments a little more carefully.

One possibility, discussed by Jensen (1989), is that the non-shared environmental component of variance reflects the operation of a large number of small and uncorrelated physical and psychological variables. The combined effects of these individually small and uncorrelated sources of variance would be similar to that of a normally distributed random variable. Of course it follows that if this were so it would be impossible to manipulate the unshared environment deliberately to produce any effect at all, improvement or not. To produce any effects the total environment would have to be controlled such that what were normally uncorrelated variables became correlated. Since the relevant variables are not known this is quite impossible. However, and this is the nub of Jensen's argument, with a random variable, some individuals, by chance, will have all influences going right for them, just as some will have them in reverse. Between such people there could be IQ differences as great as 20 or 30 points. With this interpretation it is clear that interventions are not likely to be successful.

Milwaukee project

This is particularly important in the light of the Milwaukee project (Garber, 1988) in which children at risk of mental retardation were given extensive home-enrichment programmes to prevent the expected developmental difficulties. At the age of 6 years there were gains of approximately 30 IQ points for these children, compared with controls, on the Stanford-Binet and the WISC, tests which I have described in Chapter 4. Even at 14 years of age there was a 10-point difference. At first sight these results would appear to make nonsense of our discussion of the non-shared environment and the difficulty of organising successful interventions. They appear, indeed, to cast doubt upon the biometric research.

However, Jensen (1989) takes account of the Milwaukee results without doing violence to this work. This considerable gain in IQ was not reflected in comparable gains in the two school subjects most highly loaded on *g*, reading and mathematics. This suggests that the improvement in IQ scores was brought about by essentially training for the test. In my chapter on the structure of abilities (Chapter three) I demonstrated that the variance in any test included specific variance, that is variance determined by factors that applied to the solution of that test and that test only. Thus what Jensen is arguing is that the intervention programme improved performance on the particular items and problems at the early levels of the WISC and Stanford-Binet tests. This would improve scores by improving performance on the specific factors in these tests. Such improvement would not, of course, be reflected in performance on tasks correlated with *g* and this is what was found in this study. Real gains in *g* would have produced greater improvements in reading and maths. Incidentally this phenomenon of high performance in the specific factor of an intelligence test may also be found among some members of Mensa, the society where entry depends on a high IQ score. Many Mensa members do perform at a high level, as would be expected from their high *g*, but some do not. While in some of these cases there may be reasons of social circumstances and personality, in others high *s* (specific factor) may be masquerading as *g*. In brief, the Milwaukee intervention study cannot be used as refutation of the biometric work on the unshared environment, which it must be remembered becomes more important as the children grow older. With young children shared environment is still of some significance.

As Plomin (1986) argues, and as was shown earlier in this chapter, twin studies can be used to estimate the contributions to phenotypic variance of shared and unshared environments. When this is done for samples of children, the results are similar to those of adoption studies except for the fact that there is a special twin effect. The correlation between non-identical twins is far greater than that between siblings, although the genetic similarity between the pairs is the same. Since shared environment is defined by Plomin as that which tends to make individuals similar, it must be the case that twins share their environment to a greater extent than do siblings. By such convoluted means does biometrics demonstrate a simple point that any mother or individual concerned with the upbringing of children would have known. After allowing for this special twin effect the results give rise to the same conclusions as those with adopted children.

Plomin and Daniels (1987) have examined the nature of the non-shared environment which, as we have seen, is the important influence on intelligence and, as they point out in that paper, on personality as well. As I have discussed earlier in this section, the apparently mysterious nature of this influence can be seen to be illusory if Jensen's (1989) conceptualisation of its resemblance to a random variable is grasped. That is, the effects are the result of the additive effect of a large number of small and uncorrelated influences which could cover a range of psychological, social and physiological factors, each on its own trivial. Thus the case made by Plomin and Daniels (1987), that more than one individual in the family has to be studied so that these within-family differences can be measured, does not appear well made in the light of this point. Indeed, their own preliminary finding, that scales which were able to measure differences between families were not 'sufficiently sensitive to micro environments within the family' (Plomin, 1988, p.27) runs counter to Plomin and Daniels's argument. I quote this phrase because it is another way of saying that no differences were found between the environments of children in the same family. This does call into question the notion of the non-shared environment, other than in Jensen's sense. Notice that if Jensen were correct it would be difficult, if not impossible, to pick up these differences with scales, for each would be small.

Conclusions

The conclusions that can be drawn from the biometric studies of the heritability of intelligence can be summarised without difficulty.

1. The heritability of intelligence test scores among European and White American subjects is approximately 65 per cent, meaning that this proportion of the population variance is determined by genetic factors.
2. This figure refers to the variance within the population and will vary in different populations. It does not refer to a proportion of intelligence within an individual.
3. Allowing for measurement error, around 40 per cent of the population variance is attributable to environmental factors.
4. These have been divided into two components: the shared environment, which is defined as those environmental influences which tend to make children similar, such as parental class and education, and the unshared, defined as influences tending to make children different. Parental attitudes to a particular child and the way a child responds would be examples of the unshared environment.
5. While the shared environment appears to be an important influence early in the child's life as he or she grows, the unshared environment becomes influential to such an extent that there is no

correlation between adopted children and the natural children in the family by adulthood.

6. Despite the important determination of intelligence by genetic factors it can be shown that environmental factors can create differences as large as 38 IQ points.
7. The importance of the unshared environment in the determination of intelligence tests means that intervention projects are hard to arrange, especially those within the family.
8. The non-shared environment is probably best conceptualised, as argued by Jensen, as the total effect of individually trivial and uncorrelated variables which at their extremes can create large positive or negative differences in intelligence.
9. It is quite clear from all these points that genetic factors and environmental factors are influential in the development of intelligence, as measured by intelligence tests.
10. It is, therefore, clear that performance on intelligence tests cannot measure simply skill at the tests, which would be highly unlikely to be heritable in any important degree.

National and cultural differences in intelligence

Before closing this chapter there is one final topic which needs some discussion. This concerns national and cultural differences in intelligence. This is a subject which, perhaps even more than the heritability of intelligence, excites passion and prejudice. However, in this section I want to demonstrate that clear-cut results, given all the problems of testing, are so difficult to obtain that it is necessary to remain as open minded as possible. This is because statements claiming that various national, racial or cultural groups differ widely in intelligence are seized upon and used by enemies of those groups for political purposes. There is little doubt that Blacks have suffered in this way. I think it is disingenuous for Jensen to claim that he is interested only in the scientific truth and to ignore the political and social consequences of his work. The consequences are, as he has made clear, quite illogical and are illegally derived, but this does not mean they can be ignored. This is not a plea to hide the truth, that the earth is the centre of the universe. It is simply a case of weighing up the advantages and disadvantages of hypothesising racial and national differences and their causes, especially since the evidence is not unequivocal.

Cross-cultural testing of intelligence

First I shall discuss the problems of intelligence testing in cultures different from those for which the intelligence test was originally developed. Testing in Africa is a good example and I shall draw upon my experience of intelligence testing in Ghana, where I developed a test for secondary school selection, to exemplify some of the points.

Problems in cross-cultural testing

Vernon (1969, 1979) contains a long list of the problems involved in testing intelligence cross-culturally, factors that affect the test scores such that they may not reflect the intelligence of the subjects. These include unfamiliarity with the test situation and lack of motivation. When I worked in Ghana, in some of the schools, where it was claimed a white person had not been seen before, it was difficult to get the children to take the test seriously, and to complete it. Vernon mentions anxiety, excitement and suspicion of the tester. These are common when the psychologist is of a different race. In some cultures there may be difficulties with particular types of items or materials. Lack of test sophistication in Western samples can create problems and this clearly can be even more severe in cross-cultural settings. Warburton (1951) found that the form board which requires subjects to pass objects through a matching hole was useless for the Gurkhas, who simply used force to achieve the desired end. Warburton was known as the woodman to these soldiers. Clearly this was an invalid test in this sample. Pictures or diagrams can create difficulties, which have been fully documented by Deregowski (1980). In addition to all this there may be problems in understanding the test instructions, especially if these are not in the vernacular but in English.

A problem raised by Vernon (1979) concerns the handicaps caused by poor medical care and nutrition, which can lower intelligence test scores. While these are by no means restricted to cultures other than the West they are more prevalent there. There may be other handicaps in some cultures: lack of varied perceptual and kinaesthetic experience; restricted linguistic stimulation; lack of interest in formal education within the family; restricted stimulation – no travel, television, or books; little schooling; emphasis on rote learning in school. These are the most important environmental factors which Vernon (1979) lists as likely to affect intelligence test scores. Not all cultures will show all these deficits but all are likely to limit performance in intelligence tests such that scores do not accurately reflect the potential ability of the subjects as they do in the West, where many of these handicaps are eliminated, except among a small minority.

It is possible that there are genuine genetic differences between some cultural groups such that some have special aptitudes. Such differences cannot be ruled out a priori and work by Irvine (1969) in Rhodesia certainly showed factors in Rhodesian children which had no counterpart in Great Britain, although these could well have been environmentally determined.

For all these reasons to compare cultures on an intelligence test and report the differences as reflecting differences in intelligence is highly simplistic. Any differences could simply reflect difference in some of the factors which have been discussed above.

If a test is adapted to a particular culture so that the items are meaningful (e.g. robin changed to vulture in an African setting), the tasks familiar and the instructions comprehensible and if the children or adults understand the nature of the test and its purpose so that they are trying, it probably is possible to test intelligence with some degree of accuracy. However, comparison with the results of other cultures is still dubious.

It should be pointed out, in connection with such a test, that even if all these conditions are satisfied, it is not necessarily valid. Validity must be demonstrated and this is best done by factoring the test and ensuring that its structure is similar to its Western original and by correlating the scores with external criteria. Thus an intelligence test should correlate, as we have seen, with occupational success, for example.

Does this mean that valid cross-cultural comparisons on tests are impossible? Many cross-cultural psychologists (e.g. Berry and Dasen, 1974) consider that this is indeed the case and that it is not a meaningful question to compare the intelligence of members of different cultures. This is because different qualities are valued in different cultures such that crystallised intelligence is simply not comparable (given its definition as the skills in which fluid ability is invested).

Vernon (1979) suggests that comparison between cultures might be possible if in the two groups there is equal access to education, equal freedom from physical disabilities such as those arising from malnutrition, equal familiarity with the tests and freedom from test anxiety and equal valuation of the skills involved in the test. It might be said that this, in fact, rules out most cross-cultural comparisons, certainly those between advanced and Third World countries could never meet these conditions. It is interesting to note that these criteria rule out comparison between black and white children in South Africa and many might argue that this is true of comparisons between Blacks and Whites in the USA. What is clear, however, is that comparisons on tests of intelligence between different cultures are dubious even though accurate measurement within cultures is possible if tests are properly developed.

Cattell (1971), as we have discussed in previous chapters, has argued that his culture-fair tests, which use items which are non-verbal and equally unfamiliar to all subjects regardless of culture, enable cross-cultural comparisons to be made of fluid ability. However, it does appear that familiarity with these materials affects results, a familiarity which still differs between cultures even if it is not as obvious as with verbal or information-based items, as Ortar (1963) has shown in Israel. I think that culture-fair tests are about as fair as tests could be and are effective within cultures with subjects of differing educational background. Cross-culturally their results have to be treated with great caution.

Against the background of these problems we are now in a position to scrutinise briefly the main findings in this area together with their interpretations. First I shall examine the work of Lynn and his colleagues (e.g. Lynn, 1987). Here he claims that the Mongoloids are superior to the Caucasians on Spearman's *g* and spatial ability but are lower in verbal abilities and in rate of maturation. These differences are attributed to evolutionary pressure on the Mongoloids in the extreme cold of the Ice Age, which has resulted in neurological differences, with more of the Mongoloid cortex being devoted to spatial than to verbal ability.

The data for these claims rest largely on the standardisation of various intelligence tests, especially the WISC R (the revised form) in Japan. Now as we have seen in our discussion of the problems of cross-cultural testing, true cross-cultural equivalence of tests is difficult to obtain. This is because even if all the conditions of access to education and test familiarity are met, at the item level there may be cultural differences. This makes the interpretation of mean differences difficult. For strict metric comparison it would have to be shown that the items performed identically in each culture. However, if there were real cultural differences this would not be possible. Really Rasch analysis between the cultures is the only way to make meaningful comparisons where the effect of item differences is eliminated. This is not to doubt that the factor structure of the tests is the same in the Mongoloids as in the original sample. The fact that in this culture the test is valid is not disputed. It is the absolute level of score that is the question. Huge difference would be interpretable but difference around half a standard deviation must be treated with caution. In addition to this we have to be certain, and this is difficult, that the normative sample used in Japan was a good one.

For all these reasons I would be loath to make too much of these differences in intelligence test scores even though they are in accord with other findings. They are certainly not sufficient to warrant the theorising about the effect of the Ice Age on neurological development. This is interesting speculation but not to be taken seriously. Using national changes and differences in WAIS scores over the last thirty years it would be possible, by similar arguments, to postulate that similar neurological changes had occurred in Germans and other groups – postulations which are clearly absurd. It should be noted, however, that comparisons of different racial or cultural groups within the same culture are meaningful where it can be shown that membership of that culture does not bring along with it other features which could affect the scores. Lynn *et al.* (1988) report substantially similar results for studies of children in Hong Kong where a similar difference in Mongoloid abilities was found – that is, that they are best in spatial ability and *g* and that verbal ability is low. This pattern of scores appears to be soundly established over a number of studies. The point of dispute is comparison of the mean scores with other racial groups.

Black–White differences in the USA

Jensen (1980, 1985) has made various claims concerning the differences in scores between Black and White samples in the USA. However, I shall not report these results. This is not just because there is no consensus over their interpretation, as the discussion of *Bias in Mental Testing* showed (Jensen *et al.*, 1980), but because there is no reason so to do, either theoretical or practical. In addition there is a moral dilemma involved in the reporting of this work.

There would appear to be no theoretical advance in claiming that any one race is superior to or inferior to another, given the problems and indeterminacies of racial categories (Wahlsten, 1980). Lynn, whose work on the Japanese was discussed in the previous section and who is one of the most assiduous workers in the field, can theorise only by somewhat absurd *post-hoc* means. For example it is claimed from partial correlations (*g* controlled) between WISC scales for males and females that 'during the evolution of the hominids a neurological trade-off took place twice, first when males sacrificed some of their verbal abilities ... second when mongoloids sacrificed more of their verbal abilities ...' It is unclear to me how a set of correlations on verbal and spatial ability among a Japanese and an American sample can bear on events (which are only hypothetical) taking place thousands of years ago. If this is the kind of theorising that the comparison of racial IQ scores produces then it is a fair deduction that no theoretical advances are possible.

There are many practical reasons why this work should not be reported. Black people find it offensive to be labelled less intelligent than Whites especially when (even if incorrectly) they believe the tests to be biased and that they are not given fair opportunities in a white society. Nor is it palliated by explaining that intelligence is only one variable and that Blacks are more rhythmical or better at baseball.

Furthermore, statements about the inferiority of special national or racial groups is grist for the mill of the Fascists. We saw in Germany before the Second World War objective scientists who lent their authority to the facts about races, props for genocide. Surely we should learn from the errors of Jung and Lorenz, to name but the most famous (Muller-Hill, 1988). As Claridge (1989) writes, in his review of that book, these scientists were not wicked in the ordinary sense of the word, but were willing 'to sacrifice the humanness of the person in the interests of objective truth'. I do not wish to do this, especially when it may not be true.

Finally we must answer the arguments, put forward by Jensen (1980), namely that by facing the facts (of the genetic inferiority of special groups) we can modify our education accordingly and ensure that those who need them receive special educational programmes, tailored to their ability. However, if such programmes existed and it were that easy to improve educational performance, it should be done on an individual basis of IQ scores, regardless of race. After all there is variance in the scores of all national groups and hence considerable overlap.

From this it seems to me that the only advantage in setting out the different scores on IQ tests of different racial groups is to give ammunition to those who wish to decry them. It adds nothing to theoretical understanding or to the social or educational practice.

Chapter seven

Intelligence and mental speed

The relationship of mental speed and intelligence is built into the very structure of the English language. A person who is intelligent is often described as quick-thinking or quick on the uptake. Conversely low intelligence is equated with slowness: the slow-speaking and slow-thinking country bumpkin is an obvious example. Some of our most brilliant minds are noted for the speed of their talking. Sir Isaiah Berlin springs immediately to mind and R.J. Sternberg, whose work on intelligence I shall discuss in a later chapter, is another such speaker. What this means is that, within the English-speaking world, long before the notion of psychology had ever been thought of, people had made the observation that intelligence and speed of thinking went together.

Such folk wisdom is not, necessarily, true although, in my view, it is quite wrong for psychologists to ignore human experience and live only in a world of results reported in journals. These are frequently incorrect, as future research demonstrates. Certainly linguistic assertions, of the kind I have described, are worthy of investigation. In any case it is the results of such research, relating mental speed and intelligence, which I shall examine and discuss in this chapter.

First it is necessary to consider the meaning of speed. Immediately there is an obvious difficulty. When I argued that speed was equated with intelligence in the language, this speed refers to the ability to solve complex problems or muster up counter-arguments quickly. However, there is another aspect to mental speed. This refers to speed of reaction to a stimulus. For example, in some skills just such mental speed is important. In driving, for example, it is necessary to react quickly to lights, as when the brake lights of the car ahead go on. The great psychological pioneer, Galton (1883), believed that this reaction time was positively correlated with intelligence and might be used as a measure of it, although subsequent observations of this variable failed to demonstrate the case (Wissler, 1901). Nevertheless (and this is the problem with folk wisdom) many people consider that it is unlikely, intuitively, that speed of reaction time would be related positively to intelligence, as Jensen (1982) argues. After all, intelligence is concerned with problem solving, the working through of difficult arguments. Reaction time seems entirely different, a virtually automatic process over which one can exert little control, other than, perhaps, by practice.

Most of the research which I shall examine is concerned with mental speed in the sense of reaction time because, as will be seen, it turns out that Galton was correct and that the early experimenters were mistaken in their failure to demonstrate any positive relationship. As should be obvious, from the dates of the early citations, there is more than a hundred years of work in this field although I have no intention of reviewing any but the most recent research, work which supersedes what went before.

There were many reasons why the older work failed to yield replicable results and these must be briefly mentioned in order that the more recent research can be evaluated in the light of the earlier problems. One obvious difficulty lay in the accuracy of the timing devices although, as Jensen (1982) points out, accurate apparatus was constructed. More important was the failure to realise that there was considerable variability of reaction time within each individual. To overcome this difficulty a mean is used, based on many readings in modern research. Another failing was the use of samples restricted in range, a factor which, as has been discussed, lowers correlations. These defects are easily remedied.

Another fact which must be taken into account when considering the results of reaction time studies is the nature of the experimental task. Thus it is possible that correlations with intelligence vary significantly depending upon how reaction time is measured, and I am not referring to accuracy in this instance. Whether the correlation varied or not would certainly influence any interpretation of the results. This means that different tasks need to be examined separately. Clearly in a chapter of this length it would be impossible to deal with every possible reaction time task which had ever been used. However, as the work in this area has progressed certain experimental techniques are commonly used and I shall restrict discussion to research carried out with these methods.

The Hick paradigm

First I shall describe the experimental apparatus and methods which have been given the label of the Hick paradigm from which, indeed, a considerable amount of evidence concerning speed of reaction time and intelligence has been collected. The Hick paradigm has been extensively studied by Jensen, in numerous publications, but the most complete description is to be found in Jensen (1987a).

First a few terms need to be explicated. Simple RT (reaction time) is an individual's response latency to the onset of a single stimulus. Choice RT is the response latency to any one of two or more stimuli, each calling for a different response. It is one of the oldest findings in psychology that choice RT is longer than simple RT and the difference between the two was held to be a reflection of the mental processes of discrimination and choice decision. Hick (1952) demonstrated what has become known as Hick's law, namely that choice RT increases as a linear function of the increase in the amount of information in the array (number of choices, *n*). This is the key aspect of Hick's law, for those who find the following algebraic explanation difficult. I repeat it: reaction time increases as a linear function of the number of choices (that is the amount of information) that have to be processed. That it increases is not strange. That its increase is thus regular is of considerable psychological interest.

I shall explain some of the algebra because, for a proper understanding of the work done with the Hick paradigm, this knowledge is almost essential. In this Hick paradigm information is measured in bits, bit being defined as the amount of information which will reduce uncertainty by one half. A bit, therefore, is a unit of information equivalent to the results of a choice between two equally probable alternatives. A bit is the binary logarithm (log to the base 2) of *n*, the number of choices. Thus Hick's law can be stated as: $\Delta RT = K \log_2 n$, where ΔRT is the increment in choice *RT* over simple *RT*, *K* the slope constant and *n* the degrees of choice. Thus according to Hick's law, if we know an individual's simple reaction time, his choice reaction time for any number of choices can be estimated.

Needless to say this basic formulation has been modified, as Jensen (1987a) shows, without however changing the basic logic in any way. Its modern form, as used by Jensen and most current workers in the field, is that $RT = a + b \log_2 n$ where, as before, $\log_2 n$ is the binary logarithm of the number of choices, and *a* is the intercept and *b* the slope of the regression of *RT* on $\log_2 n$.

For readers who are not certain what is meant by *a* and *b* I shall now explain these terms because they are used in interpreting the psychological meaning of the findings.

Explanation of regression

Suppose that we have two sets of scores on variables *Y* and *X*, from a sample of subjects. If we want to predict *Y* from *X* then a regression line can be drawn which enables this to be done. This line gives the best fit for predicting *Y* from *X* when a graph is drawn plotting the two scores for each individual. A similar line can be drawn for predicting *X* from *Y*. This line can be described by a regression equation:

> *Y* (predicted) = $a + bX$, where *Y* is the predicted score, *a* the intercept constant, to correct for differences in the means, *b* is the slope or regression constant which indicates the rate of change in *Y* as a function of the changes in *X* and *X* is the score on variable *X*.
>
> *a* and *b* can be better understood if we set out the formulae for their calculation.
>
> $b = r$ (the correlation of *X* with *Y*)(*S.D.Y*/*S.D.X*), where *S.D.Y* and *S.D.X* are the standard deviations of the two sets of scores.
>
> a = mean of $Y - b$(mean of *X*).

Thus given the means, standard deviations and correlation between two variables it is possible to predict one from the other. A few simple comments should be made about this regression equation and regression in general before we return to the findings concerning mental speed and intelligence.

Notice that *b*, the slope, is the same as the correlation when *X* and *Y* are expressed as standard scores, i.e. their standard deviations are 1, and means 0. When this is the case, *a*, the intercept, also = 0; *a* is called the intercept because it indicates the point on the *Y* axis that is cut by the regression line.

With all these points in mind we can now return to our discussion of the research involving Hick's law. As was indicated the best prediction of reaction time can be made from the regression equation:

> $RT = a + b \log_2 n$ where *a* is the intercept and *b* the slope of the regression of *RT* on the binary log of *n*. The intercept, *a*, is interpreted

as representing the best estimate of the total time required for all the mental processes involved: attention, sensory registration of the stimulus, transmission of the signal to the brain, reception and encoding of the stimulus, transmission of the response signal and muscle lag in response execution (Jensen, 1987a). The slope, *b*, is interpreted as the amount of time required for discrimination and choice, this time increasing at a constant rate as a function of log *n*.

As Barrett *et al.* (1989) point out, the relevance for intelligence testers to Hick's law was suggested by Eysenck (1967) who cited the work of Roth (1964). Roth reported a negative correlation of –0.39 between the regression slope, *b*, and intelligence, with *b* being defined as in our formula. This means that the more intelligent subjects showed a lower rate of increase in reaction time than did the less intelligent subjects, as the bits of information increased. On the other hand, there was no correlation between the intercept and intelligence. As Jensen (1987a) argues, these findings imply that individual differences in intelligence might reflect differences in the rate of information processing, a measure of which can be made by taking the reciprocal of the slope, *b*, expressed as bits per unit of time. While these individual differences in rate of processing are indubitably tiny, over a few years they could add up to substantial differences in cognitive skills as are found, of course, among adults. It was this study by Roth which caused Jensen to investigate the topic, findings which he reports in the 1987a paper.

There are two important features in the apparatus which Jensen has developed for investigating the relationship of Hick's law and intelligence which must be understood. Most other workers also incorporate these points. First he chooses set sizes (of lights to respond to) of 1, 2, 4 and 8, corresponding to 0, 1, 2 and 3 bits of information. This simplifies translating the equations into information-theoretic terms. In the second place movement time is separated from the RT by having subjects hold down a 'home button'. When the stimulus light comes on he uses the same finger to press the button nearest the stimulus. Subjects are told to respond as quickly as possible, the aim being to turn off the light (by the button nearest) as quickly as possible. They are informed that it is a test of the theory that reaction time is related to intelligence. This last point is important because it is likely that some subjects will be highly motivated by this fact while others may be rendered so anxious that their performance is affected. Much may depend upon their previous experiences with tests of intelligence (e.g. Sarason *et al.*, 1960).

Jensen (1987a) presents a long review of all the studies, not only those conducted in his laboratory, and I shall attempt to summarise briefly his findings. The first question that must be addressed is whether Hick's law is really a law, that is, can this neat relationship be found in all studies? This question itself is double. Does the reaction time of every individual conform to Hick's law or is there such variability that mean reaction times of groups conform (these data concealing the fact that some individuals do not obey the law)? If this is the case Hick's law is of a different order from, for example, the law of gravity.

The stability of Hick's law

The paper by Barrett *et al.* (1989) answers these questions with great clarity. They replicated the work of Jensen, paying particular care to any artefacts that might be affecting the findings. Two samples, of 40 and 46 subjects respectively, were used. To quote from the paper, they write 'not all subjects could be said to fit Hick's law'. They found in the two samples mean fits of 85 and 74 per cent compared with 88 and 94 per cent in two samples of Jensen. How can this discrepancy be accounted for?

The easy and obvious solution is some fault in the software and apparatus used by Barrett *et al.* (1989), which after careful examination they regard as unlikely. Another possible explanation is that in the British sample, some of the subjects did not really bother and were simply not paying attention to the task. If this were the case, of course we should expect to find a greater variance in reaction times in the Barrett than the Jensen data. This, however, is not the case, and Barrett *et al.* are unable to account for their findings. Nevertheless the conclusions are clear. Hick's law is not universal, although about 70 per cent of subjects appear to conform.

The relation of the Hick parameters to intelligence test scores

Jensen (1987a) averages the findings from 32 studies, which include a wide variety of samples – retarded adults, the elderly, students and high school children. He argues that these averages, although they have been corrected for attenuation due to restriction of range and the imperfect reliability of the measures, are probably low estimates of the population correlations. Since there are problems of non-fitting subjects, as has been discussed, I also include the correlations of IQ and the mean and standard deviation of reaction times, although these will have to be discussed again later in this section.

Mean RT	–32
SD RT	–48
Intercept	–25
Slope	–28

To aid interpretation of these findings, it should be remembered that the intercept is regarded as an index of the mental processes involved in the reaction time, attention, sensory registration, transmission of the signal to the brain, reception and encoding of the stimulus, transmission of the response and muscle lag in execution. The slope is interpreted as reflecting the time for decision and choice.

The first point to notice is that the Hick parameters have only low correlations with IQ tests. Thus they certainly cannot be regarded as measuring the same variable or as interchangeable measures, some kind of objective IQ test. These results have been interpreted by Jensen (1987b) as supporting the implication of a limited channel information-processing system in intelligence. Slow processing leads to an incapacity to handle complex problems where a great deal of informatiom has to be dealt with. It should be noted that these correlations would be expected to be moderate in size, around 0.4 to 0.5, since obviously capacity to reason well is separate from but dependent on efficient information processing.

Nevertheless Jensen's theoretical claims cannot be supported by the correlations which are set out above or by the findings of Barrett *et al.* (1989). In the first place they are far smaller than would be required by the theory: both Hick parameters would be expected to correlate more highly than they do. In the second place, and much more important, the theory requires that all subjects fit Hick's law. Since this is clearly not the case, this theoretical account cannot hold unless it is postulated only for some individuals, which is absurd.

Even if a theoretical account of these findings is not obvious, the correlation between variability of reaction time and IQ is substantial although, as Jensen (1987b) points out, it does not increase with set size (complexity), as it should if the theory involving limited channel capacity were correct. The correlation of mean reaction time and IQ does increase with complexity but not with all studies so that firm theoretical conclusions are again hard to draw.

Conclusions

I have scrutinised the work of Jensen on reaction time and intelligence because he has accumulated more data on this topic than any other researchers. He has also attempted to weld the results into a theory of the nature of intelligence. Nevertheless, as has been argued, the work of Barrett *et al.* (1989), who showed that not all subjects fit Hick's law and the fact that the correlations are not found in all studies, indicates that the theoretical account cannot be correct. One explanation of the problems in this area is that Jensen had originally overstated the case for the consistency of the results and that there were artefacts in the method. Thus Longstreth (1984) demonstrated that practice and order effects, response bias effects and attentional factors all influence the results and should be controlled for in the experimental design. Even more serious is the fact that Longstreth has shown that on careful examination Jensen's data are not consonant with his claims. Thus Barrett's failure to replicate them is not so extraordinary.

From this I think it can be concluded that there is a modest correlation between the Hick parameters, in those individuals who fit Hick's law, and that there is a modest correlation between speed and variability of choice reaction time and intelligence. To interpret this, however, as indicating that intelligence depends upon speed of information processing in a limited channel system is to go well beyond the evidence. In fact the paper by Barrett *et al.* (1989) contains a highly interesting analysis, a rotated component analysis of all the reaction-time variables together with the WAIS. Essentially means, standard deviations and WAIS scores loaded on separate factors. There was no support for Jensen's theoretical account.

Inspection-time (IT) and intelligence

Inspection-time is the time needed to discriminate reliably two lines of similar length briefly exposed in a tachistoscope, or on a computer screen. An auditory version is also available. IT is thought to be an index of speed of encoding stimuli, in other words it is an index of mental speed. There is nothing complex about the task, usually vertical lines which can be discriminated at one glance, hence any correlations with IQ must indicate that IQ is related to mental speed. IT studies are, therefore, another method of investigating the question discussed in the analysis of the choice reaction time data.

Brand and Deary (1982) summarise a number of studies, in which they have been involved, relating inspection-time and intelligence. Since they claim that these indicate that mental speed is an important component of intelligence and that IT will provide a culture-free measure of intelligence and a measure for infants and animals, this work deserves careful examination. This is especially true since Mackintosh (1981) has warned that the results should not be taken too literally.

I do not intend to waste too much time over the paper by Brand and Deary (1982), which is not to be relied upon. They describe in detail their latest experiment and immediately the problem is revealed. Thirteen subjects were used with a huge range in IQ: 14–54 on Raven's Matrices (max. 60) and a verbal IQ range of 59–135. Three of the thirteen were hospitalised retardates. This huge range with so small a sample is bound to maximise any correlation, quite apart from the fact that it is representative of no population. In fact IT correlated with verbal IQ –0.69 and with Matrices –0.72. Almost all the Edinburgh studies, and indeed most of the others up to this date, suffer from both the defects noted above, namely small sample size and a huge range of IQ. These two factors alone make the results dubious. Brand and Deary seem oblivious of the dangers of correlations with small *N*'s since they subdivide their samples into groups of six. With samples this size one subject with a score at the extreme end on both variables would produce a large correlation. These data, as Mackintosh (1981) suggests, are not powerful and no firm conclusions should be drawn.

Nettelbeck (1982), one of the originators of this work (Nettelbeck and Lally, 1976), did recognise the defects inherent in small samples with ranges beyond the normal population. Thus Nettelbeck (1982) used 56 students of an IQ range between 104 and 135 (median 127), an above average group. IT correlated –0.34 with the verbal IQ and –0.2 with the Raven's Matrices. Only the first of these was significant. This study indicates that mental speed, as measured by IT, could not possibly be used as an alternative measure of intelligence and that it is only modestly related to intelligence in a sample not distorted by retardates whose IT performance is always markedly poor, a finding incidentally which casts doubt on the validity of IT as a measure of the speed of mental processes. It could be more an index of qualitative differences in attentional style.

Generally, it may be concluded, when samples of a reasonable size and excluding subjects of very low intelligence are used, there are only modest correlations between measures of IT and intelligence, far too small for practical measurement or specific theorising.

Finally I should like to mention a small study which I carried out in Exeter (Cooper *et al.*, 1986). In order to understand better the correlation between IT and IQ it seemed sensible to investigate its correlations with the main primary ability factors, which have been described in a previous chapter. To this end the correlation between IT, measured on a microcomputer, and the primary ability factors in the Cattell Comprehensive Ability Battery (Hakstian and Cattell, 1976) was investigated in a student sample of 20 subjects. Here it was found that, not unsurprisingly, IT correlated best with the factors of perceptual speed and visualisation, rather than the primary factors related to fluid and crystallised ability. This suggests that IT may tap perceptual processes rather than intelligence, a notion with face validity.

Conclusions

The conclusions from the work on IT and intelligence are clear and almost the opposite of those claimed by Brand and Deary. IT cannot be used as a measure of intelligence and its relation to IQ is small, even if statistically significant. No great theoretical superstructure should be built on this research.

The findings concerning choice-reaction time and intelligence are somewhat similar. There is a modest correlation with variability of reaction time and in some samples with mean reaction time. The Hick parameters correlate in some samples with IQ but not all subjects fit Hick's law. There is no constant increase in the correlation of reaction time variables with IQ as tasks become more complex. The implication of speed of processing, in a limited channel system, with IQ is not supported by these findings.

Chapter eight

Cognitive processing and intelligence

So far in this book, it has been shown that two factors, fluid and crystallised ability, can be extracted reliably from a battery of ability tests and that they correspond to most concepts of intelligence. It was shown that there is a considerable genetic component in the population variance of this factor which is powerful in occupational and educational psychology. It was further demonstrated that, despite certain claims, these factors could not be identified with mental speed. Nevertheless all this work is strong confirmation of the utility, not to say necessity, of the notion of intelligence, factorially defined.

There can be two objections to the factor analytic description of abilities, even among those who are sympathetic to the psychometric approach. One was raised by Cattell (1971), who argued that a major difficulty with the psychometric work in human abilities was its divorce from cognitive psychology. Attempts to theorise about the nature of abilities were hindered by this fault. Thus, to give an example, even if it is admitted that intelligence plays a part in working out calculus, it does not tell us what goes on. It is far too general and imprecise a construct.

Cattell (1971) tried to remedy this defect but, impressive though his attempt was, it still fell outside the mainstream of cognitive psychology. Hunt (e.g. 1976) has made this point and has carried out considerable research in which he attempts to elucidate factor analytic abilities, especially intelligence and verbal ability, by investigating their underlying cognitive process. In so doing he raises a second objection, namely that, within any two individuals, the same performance may not be a linear function of a fixed set of cognitive processes. This means, of course, that the psychology of individual differences cannot be recovered from data analyses which use linear models. Factor analysis is, as was discussed in Chapter 2, just such a linear model.

I shall take the second point first. If it were true that non-linear processes prevailed in problem solving then the factor analysis of abilities would have been unable to extract meaningful factors from the study of individual differences because no regularities would underlie the differences. However, as has been seen, this is manifestly not the case. It is not possible to sweep away 70 years of factor analysis on the basis of a few small experiments with necessarily tiny samples. The fact is, judging from correlations with external criteria, ability factors are of practical utility and make theoretical sense. As Cattell (1978) argues, in science it is best to make the simplest assumptions, in this case linear models, until the data show that they are wrong. This they have failed to do. Thus I think that this second point is mistaken. The structure of abilities has been extracted from the study of individual differences by linear factor analysis.

Now to the first point, namely that the intelligence factors do not specify mental processes and that psychometrics is divorced from cognitive psychology. First I shall discuss the divorce of psychometrics and experimental psychology, a state which has been changing rapidly in the last ten years so that now, indeed, there is something of a shot-gun marriage between them.

Most writers in the field assume that this divorce is a bad thing. This is hardly surprising since cognitive psychology with its close relationship to the holy grail of modern science, the computer, is the dominant strand of psychology. M.W. Eysenck (1984) claims that about 80 per cent of published work falls into this category. However, as I have pointed out elsewhere, certain aspects of cognitive psychology are far from intellectually powerful, especially those which deal with memory models (Kline, 1988). There it was shown that many of these were examples of what has been referred to as non-contingent statements (Smedslund, 1978), i.e. they must be true by virtue of the language in which they are couched. In one model it is demonstrated that recall requires more information than does recognition. Yet this is built into the definition of the words. No evidence is needed. Retrieval depends upon information stored in the memory. These are banalities. It would be inconceivable if this were not the case. Thus it is by no means an unarguable case that psychometrics should join hands with cognitive psychology.

Actually I think it should and for reasons that are relevant to the second point, namely that generally psychometric factors do not specify processes, do not describe precisely what the mental operations are, while cognitive psychology attempts to do so. In principle this is a useful distinction. However, examination of the list of primary factors and even some of the secondaries does not make this distinction as obvious as it first appears. For example, there are primary memory factors such as rote memory, the ability to remember pairs of stimuli. This must be similar to processes involving short-term memory. Nevertheless it is certainly true that if the relationship of ability factors to cognitive processes were worked out there would be useful knowledge, both in understanding factors and the processes themselves.

Since the late 1970s psychometrics has definitely adopted the view that a close relationship with cognitive psychology is the way to go. Hunt and his colleagues have explored verbal ability and intelligence and Sternberg (1985a), at Yale, had developed a theory of intelligence and ability based upon his studies of processes. Carroll (e.g. 1983) has also done much to explicate the meaning of factors in terms of cognitive processes.

There has been a huge output of this work since it offered the hope that the ability factors could be better understood than from factor loadings with other tests and external criteria. This makes it impossible, in a single chapter (of finite length) to review it all. Rather I shall describe what I believe to be the most psychologically significant findings, from the viewpoint of the explication of *g*. First I shall examine the work of Carroll, which is an a priori analysis of the relation of factors to cognitive processes and which is valuable as an aid to research design in this field. I shall then scrutinise the work of Sternberg and his colleagues at Yale on the componential analysis of abilities, research that is at present the most influential of all in the field of intelligence. Next I shall examine the research undertaken by Hunt and colleagues who were pioneers in the cognitive psychological approach to intelligence. Finally I shall attempt to draw together some conclusions from all these different approaches.

The work of Carroll

Carroll is currently engaged upon a huge study of all the factors that have emerged from the factor analysis of abilities. This is an attempt to produce a definitive list (although not a complete one) by rotating all factors to simple structure, as I described in Chapter 2. In addition to this, and almost as part of this work, Carroll (1980) carried out an extensive study of the cognitive processes which have been cited in the experimental psychology of cognition (Posner, 1978; S. Sternberg, 1960, are good examples). To this end he subjected the correlations between the elementary cognitive tasks (ECTs), from which cognitive processes are inferred, to factor analysis despite technical problems arising from the small samples and rather few variables used in this type of experimental work. These factors, and their possible relationships with the ability factors derived from psychometric tests, have been described in minute detail by Carroll (1980).

Before I discuss these cognitive process factors, I shall describe a few typical ECTs so that the empirical basis of these factors can be understood. First I shall define an ECT. It is a task, the completion of which requires a relatively small number of mental processes or operations. Clearly this is so broad that there will be a huge number of ECTs so it is convenient to group them in some way. Carroll divided ECTs into eight categories, although it should be noted these were a priori divisions, not empirically defined. These were: perceptual apprehension; reaction time and movement; evaluation and decision; stimulus matching or comparison; naming, reading or association; episodic memory readout; analogical reasoning; algorithmic manipulation. It should be pointed out here that the category of analogical reasoning is particularly interesting. This is because analogies are part of the psychometric WISC and WAIS tests which I have described. Even more importantly, the solution of analogies was the basis of the work by R.J. Sternberg on the componential analysis of abilities, work which I shall discuss in a later section of this chapter. Actually, as Carroll admits, this category is different from the others since the solution of analogies would appear to involve some of the other categories. Hence the term 'simple' may not be accurate.

Another way of looking at the tasks is in terms of underlying cognitive processes and Carroll lists nine of these which will be familiar to those versed in experimental psychology. These are: monitoring, attention, apprehension, perceptual integration, comparison, co-representation (as used, for example, in foreign languages) co-representation retrieval, transformation (as in mental rotation), response execution.

I shall describe a few examples of ECTs, as found in the literature of experimental psychology.

1. *Perceptual Closure Tasks* These require subjects to interpret or recognise ambiguous stimuli. In *Degraded Words* words are

presented on a computer screen, degraded to various extents by switching off different numbers of pixels. Depending upon the precise set up and purpose of the task, response latencies and number correct can be used as scores. The effects of increasing the information presented can also be examined. This ECT is expected to measure perceptual closure.

2. *Ambiguities* In this ECT, phrases are presented and the subject has to judge them true or false. It is, therefore, a form of lexical decision task. In the study by Kline *et al.* (1986) three categories of phrase were used: true (e.g. 'a dog has legs'), false ('a book has legs') and ambiguous ('a plane is a tree'). Each of the ambiguous phrases was true if the less common meaning of the noun were apprehended and false if the more common meaning was taken. This ECT was designed to measure speed of access into internal lexicons. If a subject was slow to access his lexicon then the less common meaning would not interfere with the response 'false' to the ambiguous stimuli. Those with rapid access would either respond 'true' or have delayed response latencies as the conflict was resolved. This ECT is thought to tap evaluation and decision.
3. *Reaction Time* In this ECT the measure is latency of response in a choice-reaction task when the subject has been previously primed as to which of the choices is likely to occur. This latency is compared to that when there has been no priming. Latencies to rare events can also be measured.
4. *Reaction Time* This is a similar ECT to the one above. Here, however, the effect of presenting a stimulus which requires a different response just before the target stimulus is measured. Both these reaction time ECTs are suited to the study of the divergent factor, studied especially by Guilford, as well as general intelligence. For example in flexible subjects the effects of presenting competitive stimuli would be expected to be less than in convergent individuals.
5. *Shape Comparison* This is an ECT which has received minute and detailed scrutiny in cognitive psychology, especially by S. Sternberg (1975) and Posner (1978). Here subjects are presented with either two shapes, two names of shapes, or a shape and a name, the task being to decide whether the members of a pair refer to the same shape. The reaction time to shape and name should be longer than the others because of the extra processing time required for the access to the name code.

These typical ECTs give an idea, it is hoped, of the data on which the attempt to link ability factors with cognitive processes is based.

I shall not list all the factors which were discovered but give examples of some of these factors from various cognitive domains. Carroll (1983) contains a valuable summary of some of this work and readers must be referred there for further details. In the perceptual domain speed of closure, perceptual speed and spatial accuracy were found. In the domain of reaction time (a field which was discussed in Chapter 7) the Hick slope emerged as a factor as did simple reaction time and hand movement speed. In a domain which Carroll labels accuracy of information processing a factor was found concerned with accuracy in semantic processing. In the domain of response to orthography (the one including all tasks which require response to letters and words, for example), an encoding factor was found together with a depth of processing factor. In the domain of memory there were a number of factors, including three memory span factors and a memory for complex events.

Before the psychological implications of these factors for understanding the nature of intelligence can be understood a few points need to be borne in mind. First, these domains are intuitive categories not empirically based on hierarchical factor analysis. Second, the naming of these factors is tentative because few if any have been used in validating studies. Furthermore, few studies used the same sets of variables which makes resulting factors difficult to compare and thus identify. Finally, although Carroll does attempt to indicate how they relate to the psychometric ability factors this again is a priori rather than empirical. In the ten years since the monograph was written there have been few studies in which the nature of the processing factors has been elucidated by joint factor analysis with the older psychometric factors, so that the interesting arguments developed by Carroll are still the best guide to understanding how psychometric and ECT factors relate to each other.

Interpretation of ECT factors

In our chapters on factor analysis and on the ability factors that have emerged from its application to psychometric ability tests, the distinction between general and group factors was made as it was between first and second order factors. In each case it was a question of breadth. Second order factors accounted for correlations between primary factors and were necessarily more general. Group factors load, on several tests, general factors on all of them but this is a somewhat arbitrary distinction, depending on the variables in the battery.

The relevance of this to understanding the nature of the ECT factors is that ECT factors tend to be highly specific or very narrow factors, which is hardly surprising given that each is concerned, almost by definition, with a particular cognitive process, many of which would be expected to contribute to a fairly broad ability. A few examples will clarify this point.

For example in the perceptual domain both speed of closure and perceptual speed factors have been found among ECTs. These, of course, are also in the list of primary ability factors. Here, then, the narrow primaries and the ECTs appear actually to be the same. Again in the perceptual domain, spatial speed and spatial accuracy are found. In the sphere of ability factors we find spatial orientation and spatial visualisation. Now, as Carroll (1983) argues, it is highly likely that both these ECTs are part of both or either of these factors. In this instance the ECTs are narrower than the primary factors and, as I have argued before, this is the more usual case.

The ability factor of perceptual speed is particularly interesting from the viewpoint of the relationship between ECTs and ability factors. This is, of course, because, as the description of the ECTs themselves made clear, we would expect many ECTs to load on this factor. That is, ECTs would appear to measure the fine detail of perceptual speed. Examination of the list of ECT factors supports this argument. For not only does one of the ECT factors appear to be highly similar to the *P* factor, two of them seem to be aspects or components of *P*: speed of mental comparisons and the speed of visual and memory search. It should be noted, at this point, that these results with the *P* factor indicate a problem with ECTs, namely that some are more simple than others. As Kline *et al.* (1986) found, careful analysis of some of the ECTs showed that they measured, at least putatively, more than one cognitive process.

In the field of memory there is also considerable overlap between the ability factors and the ECTs and again this is not surprising, given the nature of the ECTs. There are ECT factors which are highly similar to the primary factors of memory span and associative memory.

From this discussion I think that there can be little doubt that ECT factors are not intrinsically different from primary ability factors. They would appear to be narrow factors, tapping mental processes which are involved in the primary abilities, although in some cases, where the particular ECT is not so simple as to tap one process only, the ECT factor may be identical with the primary factor. Clearly work that relates ECTs with ability factors empirically will be of great value in elucidating the nature not only of the primary factors but also of the two *g* factors of intelligence. This work, however, cannot be examined until I have described and discussed the work of Sternberg, whose triarchic theory of intelligence has become the most well-known account of intellectual abilities.

The work of Sternberg

In 1977 Sternberg published in full detail his componential analysis of intelligence and since then there has issued from his Yale laboratory a seemingly ceaseless flood of research and writing in which it has been developed and elaborated. In the 1985 version *Beyond IQ* he cites more than 70 references to his work and as I write this in 1989 this count has doubtless doubled. Heraclitus said that you can never walk in the same river twice and as I contemplate summarising all this work, I know the feeling.

I shall make no attempt to give an account of the whole theory because, in fact, Sternberg has embraced under intelligence almost all human behaviour and has gone well beyond what most people would mean by the term. In a somewhat pragmatic and American concept, being intelligent is seen as being successful. Thus Sternberg's (1985a) list of deficits in intelligence tests includes the facts that there are no measures of adaptation to, selection of and shaping of real-world environments; no dealing with novel tasks and situations; and no metacomponential planning and decision making, by which Sternberg refers to the strategies used in the solution of problems. These faults are sufficient to indicate that Sternberg's notion of intelligence is extremely wide. It also appears to be concerned with aspects of motivation and includes, further, social competence. Nevertheless Sternberg's work has been powerful in elucidating the nature of intelligence as defined by tests of intelligence and it is on this aspect of his work that I shall concentrate.

Components

In the 1977 book, to which I have referred, Sternberg attempted to break down into its constituent mental processes the solution of analogies, which are regarded as one of the best measures of the *g* factor. To do this he developed an ingenious task known as the People Pieces. These are schematic figures which possess four salient and bivalent features: sex, height, colour and girth. Any four of these figures can be used to form the terms of an analogy and the subject has to decide whether the analogy is right or wrong. By varying the number of features that change within the terms, the difficulty of the analogy can be manipulated. In addition to this, precueing can be used so that subjects can see some of the terms before the whole analogy is shown. This means that certain of the component processes have been executed before the response time for the solution is measured. Hence estimates of the time taken for these componential processes can be computed. Responses to the People Pieces were the original data base of Sternberg's componential analysis.

In this model of analogies five components were identified – encoding, inference, mapping, application and response. Encoding involves translating a stimulus into a mental representation, for example that this is a fat, short, red, male. Inference is to find the rule relating the first and second term of the analogy while mapping finds the rule that relates the two halves. Application is applying the inferred relation of the first half to the second half to arrive at the solution. Response is communicating the response, which depends on the experimental set-up, speaking, pushing a button, or filling in an answer sheet.

Sternberg (1986) in a summary paper in *American Scientist* claims that higher scorers on standard tests of intelligence are faster at inferring mapping and applying relations, as well as responding which is, perhaps, a component of a different type. Sternberg and Gardner (1983) found that component scores derived from three different tasks (analogies, series completion and classifications), which are all described in Chapter 4 on intelligence tests, with three different contents (schematic picture, verbal and geometric), had an average correlation with intelligence test scores of around –0.61. The correlation is negative, of course, because speed of reaction time is the measure from which the componential scores are derived. Although this correlation is impressive I must point out that it is not that surprising given that analogies, series completions and classifications are all important measures of the *g* factor. Does it mean anything more than that componential scores derived from a set of tests correlate highly with success at those tests? It probably does because these components are mental operations, even if they are not precisely defined, and these processes are thus implicated in intelligence.

However, and this indicates that the relation of components to intelligence test scores is not entirely tautologous, it is interesting to note that the high scorers on intelligence tests were slower at encoding, presumably to ensure that they do so in detail so that in attempting to solve the problem they do not have to refer again to the original stimulus or recode it in some different manner. In fact it appears that in a typical verbal analogy about half the solution time is taken up in encoding. From these results Sternberg is prepared to claim that the general factor in human intelligence can be understood in terms of performance processes. Thus psychometric *g* has been explicated.

This is the aspect of Sternberg's work that is most relevant to understanding the *g* factor or factors. However, scrutiny of this work indicates that it is not as meaningful as it first appears, a criticism that can often be levelled at cognitive psychology, as has been explicated by Kline (1988). In part, at least, Sternberg is aware of this which is why he has turned his attention to other aspects of intelligent behaviour, as will be discussed later in this section. However, let us examine the componential analysis of *g*. According to this then the solution of a problem involves encoding, mapping, inference, application and response. I believe that this is an example of what Smedslund (1978) has called pseudo-empirical psychology, that is, it must be true from the nature of the language. If it were untrue the language would be meaningless.

For the sake of clarity I shall set out an analogy, A:B as C:D. I shall now examine the components used in solving it. First we encode the analogy, make a mental representation of it. This must be so. It is impossible to envisage any kind of problem solving without mental representation. A blind person (*pace* blindsight) could never solve a visually presented analogy or a deaf person one in the auditory mode. Thus to postulate encoding in problem solving is banal, if not vacuous. In defence of componential analysis, it can be said that it has been found that highly intelligent people spend longer in encoding than do the less intelligent. This is a genuinely empirical point.

Then we find that inference and mapping are involved, inferring the rules between the terms and the halves. Again this must be so. An analogy, by definition, could not be solved unless these rules were worked out. No experimental work is needed to demonstrate this. What has been discovered is that the faster one is at these processes the more intelligent one is. Given the well-known relation between speed and power in the solution of problems, this empirical point is of no great interest and could have been inferred, since these processes must be part of the problem solution. Similar arguments apply to application and response. This last process is truly banal, the time taken to make the response. It needs psychology to discover that process.

In brief it can be seen that these components are non-contingent in that they are built into the meaning of analogy, a reasoning process that was mentioned by Aristotle. Thus the only empirical information in the Sternberg analysis, that could not have been inferred from the fact that power and speed are correlated in tests, is that intelligent subjects spend more time encoding the information, at least in these laboratory tasks. It is certainly not true to say that componential analysis has powerfully explicated the nature of intelligence.

An obvious question concerning these performance components concerns their generality. Pellegrino (1985) in his study of inductive reasoning shows that they apply in classification and in series completion items as well, as careful thinking about the nature of these items (fully described in Chapter 4 on intelligence tests) suggests. This is not surprising since inductive reasoning is highly similar to the definition of *g* and all these items load highly on the *g* factor. However, May *et al.* (1987) showed that these components played no part in flexible thinking as measured by divergent thinking tests or ratings of flexibility. This supports my claim that these are non-contingent concepts entailed in the notion of analogical and other similar reasoning. Analysis of flexible thinking suggests rather that performance acquisition components (Sternberg, 1985a) would be more relevant but these were not used. These components are discussed in the following section.

The triarchic theory

In the triarchic theory of intelligence, which I shall examine, since it is argued there that the factorial *g* is too narrow a concept of intelligence, Sternberg (1985a) involves three aspects of information processing in intelligence although this use of the term information processing is a broad extension of its meaning. These aspects concern the mechanisms of intelligent functioning (of which performance components are a part); the point of practice in a task when these mechanisms play their part; and the relation of intelligence and the external world (shaping, adapting to or selecting the environment). These are the questions with which the triarchic theory is concerned. This last aspect is a long way from the conventional meaning of information processing. Indeed so is the second aspect which states that intelligence is only displayed where there is a novel task or where the materials of the problem are so well known that handling them is virtually automatic. Again it should be noted that this must be true since by definition it would be difficult to act intelligently as defined (educing relationships) where the answer to the problem was known, where no reasoning was involved or where the materials were so unfamiliar that no reasoning could take place. This is the well-known principle behind the construction of culture-fair tests, as has been discussed in Chapter 4 on intelligence tests. Thus the only aspect of the triarchic theory which is really relevant to our analysis of intelligence is the first aspect, concerned with its underlying mechanisms.

The mechanisms which Sternberg (1985b) proposes are: meta-components, such as deciding on the nature of the problem and selecting a strategy for its solution; performance components, which have been dealt with under componential analysis, and knowledge acquisition components, the processes used in the acquisition of new information. Sternberg argues that mechanisms from all these three classes are involved in intelligence as measured by IQ tests.

The performance components have been discussed and I have shown that these are non-contingent concepts. Their truth is entailed in the language, thus making experiments pointless. However, it should be noted that their descriptions are not precise. For example, performance components are described by Sternberg (1985b, p.18) as non-executive processes used in actually executing a strategy. I cannot comprehend what this might mean.

I shall now turn, more briefly, to the other classes of mechanism. First I shall examine the knowledge-acquisition components. Vocabulary is the best predictor of IQ, if one brief test is to be used (Vernon, 1960; Jensen, 1980). This is presumably because it reflects the ability to acquire meanings from natural contexts which, in turn, reflects the ability to educe correlates. Thus the study of how the meanings of words are acquired was used by Sternberg and Powell (1983) in their investigation of these knowledge-acquisition components. In this research subjects were given passages of English in which non-English words were embedded and the meanings of these had to be worked out from the context.

Sternberg (1985b) argues that there are three critical processes in the elucidation of meaning: selective encoding, which involves figuring out what information is relevant in learning, the meaning of a new word; selective combination, which involves putting together in a meaningful way what has been selectively encoded; finally selective comparison is used to relate the new to the old information, and thus work out the meaning. Furthermore he argues that these three processes are not limited to the acquisition of vocabulary but are involved in other kinds of learning, especially insightful learning, and that insights can be of three kinds corresponding to the three mechanisms.

These are highly interesting claims and I shall discuss Sternberg's examples in some detail, because they again force me to the conclusion that these are non-contingent concepts. However, before I do this a more general point needs to be discussed. Sternberg's language is so imprecise here that I believe that his conclusions are false. The ability to work out the meaning of a word is not identical to acquiring its meaning. Thus although I worked out the meaning of the artificial word in the example in his text I can no longer remember it. Furthermore studies of human learning indicate that practice plays a part in acquisition. Thus while it is true that this ability to figure out the meaning is important in language acquisition there are many other factors. While this may be a small and somewhat trivial point, the next is not. Insightful learning is claimed to involve the same components, and one of his examples is Fleming's discovery of penicillin. My point is simple. The use of the word learning in this context seems quite wrong. Indeed it appears that no learning, just good reasoning, was involved. Thus to call these mechanisms knowledge acquisition components seems misleading. Indeed in a later section of this paper Sternberg refers to gifted children having these insights spontaneously, which renders the notion of learning incoherent.

In support of his notion that selective encoding, selective combination and selective comparison underlie insights Sternberg examines three famous examples, Fleming, Darwin and Kekule, although he goes on to describe some of his own empirical research. Fleming's discovery is said to be an example of selective encoding, in that he 'zeroed in' on the relevant material, the bacteria destroyed by the mould. There are two points here. He selectively encoded the information, it is true, but the fact that it was relevant to the discovery was fortuitous. Selective encoding is a general mechanism. The fact that it led, in this case, to insight by no means demonstrates that it is an important aspect of insight. I might selectively encode an 'irrelevant' aspect and that would actively prevent discovery. The important thing is that he saw the relevance. This is not identical with selective encoding which, after all, takes place in working out the meaning of words.

Even more important, however, is the further point that this concept is non-contingent and thus somewhat tautologous. Let us ask ourselves what happened when Fleming saw the mould. He must have paid attention to the dead bacteria. If he had not, obviously there would have been no discovery. It is therefore banal to say that a mechanism of selective encoding was used. The fact that he selected one part rather than another entails the mechanism, if we are going to describe what he did in terms of mechanisms. No experiment is required to demonstrate this point.

Selective combination is said to underlie Darwin's theory of evolution. As Sternberg argues, what distinguished Darwin from other biologists at this time was the fact that he selectively combined the information to form coherent theory. Again this is bound to be true. If a theory is formed by combining elements of an array of information, selective combination (or a homonym) must be the mechanism involved. Suppose I were to argue that Darwin absolutely did not combine selective aspects of information to create the theory. Clearly this would be senseless, just as if, in the case above, I were to claim that Fleming ignored the bacteria on the plate. Non-contingent concepts, of course, must be true, hence it is hardly surprising that experimental evidence can be gathered in their support. However, because they must be true the gathering of such evidence is pointless. Again, it must be said, selective combination does not, necessarily, lead to insight.

Selective comparison, as a basis of insight, is exemplified by Kekule in his discovery of the Benzene ring. Kekule, it is claimed in almost every text on creativity, discovered the structure of the Benzene molecule, having dreamed of a curled snake, tail in mouth. Using the mechanism of selective comparison he saw the analogy between the reptilian image and the molecule. All the same arguments apply. Selective comparisons can be made which are downright misleading and hold up progress in science. For an example, by free association I could conclude that the slow worm, from its absence of legs, flickering tongue and speed of motion, was a snake. More important is the argument that, as was the case with the other mechanisms, selective comparison is non-contingent. Given that Kekule saw the solution to his problem in the shape of the snake, selective comparison is the only mechanism that could be put forward. It is senseless to argue, for example, that he solved the problem by ignoring the snake.

As was the case, therefore, with the performance components, these knowledge-acquisition components are necessarily true and thus cannot be confirmed by empirical investigation. However, these components may not be correctly named as concerned with knowledge acquisition since in all these instances they seemed to be concerned with the solution of problems, which, ultimately, it is true, did give rise to knowledge.

It must be realised, however, that empirical research into non-contingent concepts, such as these knowledge and performance components, is not out of place. It might be possible to explore them, find out, for example, the conditions which improved their efficacy or when they were not correctly employed. However, what is not useful, as I have argued, is to cite or carry out empirical research in support of their postulation. This is an important point because Sternberg (1986) argues that identification of these cognitive processes provides us with an understanding of intelligence that cannot be derived from intelligence test scores alone. This is so but it does not require experiments to identify them.

I shall now scrutinise the third set of components in the triarchic theory. These are metacomponents, metacognitive processes. Sternberg (1986) claims that identification of the cognitive processes is not enough. The metacognitive processes that drive the cognitive ones must be understood. Would that Aristotle had not invented the term metaphysics, which means simply after physics. By the barbarous the term meta has been endowed with sublime qualities: they long to use it. Metacognitive processes direct and receive feedback from cognitive processes. Although vaguely formulated, the notion of directing feedback is perilously vague, unless its destination is clear; a number of such processes have been proposed, although there is less agreement here among different investigators. Sternberg includes the following: recognising that a problem exists; defining the problem; selecting a set of cognitive processes to solve the problem; constructing a strategy for the solution of the problem; selecting a mental representation on which the processes and strategy will operate; allocating both external and internal resources to the problem and monitoring one's solution to the problem.

Some of these components are, without question, non-contingent, although others require careful scrutiny. Recognising a problem exists falls into the first category. Actually I doubt whether any meaning whatever could be attached to a component of this description. First, it is inconceivable that a problem might be solved without first recognising its existence, other than by chance, which would never be repeated. Thus, in this sense, this metacomponent must exist. However, the existence of a problem implies recognition. Without it there is no problem. This mechanism is non-contingent. Defining a problem is similarly tautologous. How could a problem be solved if it is not defined? It is impossible to find any meaning to a claim that an undefined problem might be solved. It follows, therefore, that this is another non-contingent concept. Again it must be emphasised that these non-contingent concepts can be subjected to some empirical investigation. However, their postulation does not demand experiment, nor is their postulation noteworthy.

I shall not go laboriously through all the metacomponents, for reasons of space and wearisome repetition, but I shall scrutinise one of the more important, cognitive representations, since Sternberg (1986) claims that more intelligent problem solvers differ in the way that they represent problems from the less intelligent. For example younger children represent stimuli holistically while older children and adults represent them analytically. Verbal individuals tend to use verbal descriptions while spatial subjects used spatial representations, a finding described as highly interesting. First, although this is said to be a meta-component, it is somewhat similar to encoding, the performance component. Again it must be a non-contingent concept since any kind of mental problem solving must entail mental representation. However, in this instance the empirical research has revealed information about the nature of the mental representations. For example, expert and novice chess players differ in their representations: experts could draw on thousands of chess patterns stored in long-term memory, whereas novices have relatively few. Thus chess players were, essentially, making selective comparisons between the board and thousands of representations stored in memory. Thus knowledge is important in problem solving, and not only in chess. Sternberg (1986) writes '. . . the amount of knowledge makes a crucial difference to performance in a variety of fields, although in order for the knowledge to be used as needed it must be represented in a way that makes it retrievable' (*sic*). This last sentence is surprising since it is hard to see how the knowledge could make a difference if it could not be retrieved. Nor is it useful to think of knowledge (of any kind) as something that cannot be retrieved. By definition we cannot know what cannot be retrieved and it makes no sense to think of knowledge that we do not know. That is why I am forced to disagree with Sternberg's (1986) claim that these methods are keys to unlocking the mysteries of the mind. This seems a grotesque exaggeration, given the nature of the concepts involved.

I do not want to say any more about the work of Sternberg. Despite its conceptual inclarity, there is a valuable core of work in which mental processes can be explicated, often by ingenious experiments. However, the concepts in the main are not empirically supportable but are non-contingent, inevitable notions given the language in which we talk of problem solving and thinking. If this were frankly admitted some of the banalities and tautologies of the theory could be removed. As it stands, however, triarchic theory needs severe revision.

The work of Hunt

Finally, before concluding this chapter, I shall discuss some of the work of Hunt and his colleagues who were among the first to attempt to study ability factors, mainly verbal ability, by using the methods of cognitive psychology. Hunt (1976, 1978) summarised much of the work, which essentially consists of showing that speed of lexical access is an important component in verbal ability and in intelligence. Indeed there is a correlation of about 0.3 between IQ scores and speed of lexical access. Hunt used a model of memory which was current at that time and which included short-term memory, intermediate memory, and long-term memory, with various buffer stores. It was, in essence, a simple computer model of memory. In this model information is passed up (via buffers) towards short-term memory in a series of progressively higher order codes. These codes are based on a match of sensory input with patterns in long-term memory. When information in a buffer is recognised, the appropriate pattern name is placed in the next higher buffer, until finally all information is placed in short-term memory when it becomes conscious. Active information processing takes place in short-term memory and in the intermediate memory. Short-term memory contains an echoic memory of recent stimuli, intermediate memory, and information structure capturing the meaning of ongoing events. The short-term memory lasts seconds, the intermediate memory lasts minutes or hours, and the long-term memory is permanent.

Although this is an old model, as M.W. Eysenck (1984) demonstrates it does contain most of the critical features of more modern models, at least from the viewpoint of lexical access. Hunt argued that differences in verbal intelligence could be understood largely in terms of differences in speed of access to lexical information in long-term memory, this lexical information being retrieved through the system of buffers and stores which was described in the previous paragraph. In the ingenious experiments which Hunt and his colleagues use, lexical access time is distinguished from speed of response time with remarkable simplicity. The time taken to say whether two letters are physically identical or a name match is compared. This difference is regarded as an index of speed of lexical access which is not required for the first task.

The advantage conferred by rapid access to the lexicon is that it enables more time to be spent on other mental operations. In addition Hunt has demonstrated that in high scorers on verbal intelligence tests there is rapid manipulation of data in short-term memory, a better capacity to retain order of entry of information in short-term memory, and a better organisation of data in long-term memory. However, all this work is based on highly artificial laboratory tasks and Hunt admits that with meaningful materials findings could be different.

It is generally agreed that the major finding in this work is the correlation between speed of lexical access and intelligence. Although the correlation is not large, it is reliable. What does it imply about the nature of intelligence? First, it is too small to argue that it is in any way causal: it cannot be that people are intelligent because they have rapid access. Nor could the opposite be true, that they had rapid access because they were intelligent. I think the meaning of this finding is relatively simple. Rapid access confers a slight advantage for carrying out mental operations and this is reflected in the correlation with IQ. However, I would be surprised if that were the sole explanation of this correlation. I would expect that speed of access reflects more efficient neural functioning in general, of the kind discussed in the previous chapter where small correlations were found between IQ and mental speed and lack of variability as measured in choice reaction tasks in the Hick paradigm.

Conclusions

As was mentioned at the beginning of this chapter, Hunt (1978) especially and Sternberg (1977) claimed that the experimental cognitive psychological approach to the understanding of intelligence was superior to the old factor analytic method. Sternberg (1985a) is now less confident, claiming rather that they are a complement to the factorial work. Nevertheless even here he is cautious, arguing that it is difficult to reduce intelligence to a series of processes or components which are derived from such simple tasks, unless, of course, the multiple correlations were unity. Simple letter comparisons, as Hunt uses, or even the People Pieces analogies, are qualitatively different from the complex problem solving required in the reality of life. Of course, this objection has been laid at the door of intelligence tests. This is also a real objection, although intelligence test items are not as simple as the tasks of cognitive psychology which are deliberately so designed. In addition they do have substantial correlations with real-life criteria.

In addition I think there are some further difficulties with the cognitive psychological approach. The first concerns the implicit models which underlie them. These models, for example the memory model used by Hunt, obviously dictate the kinds of processes involved. Now these models are nothing more than models. They are certainly imperfect since there is no agreement concerning which is the best and predictions derived from them are not always supported. In addition, as I pointed out in the case of Sternberg's work, the components were non-contingent entailed by the language. There is nothing wrong with such concepts *per se* although it is pointless to seek empirical evidence in their support. Empirical work should concentrate on their explication and description.

From this it is clear that as yet the contribution of the studies of cognitive processing to understanding intelligence has not been large. This is undoubtedly why, in the triarchic theory of intelligence, Sternberg plays down this aspect of his work and introduces a far broader notion of intelligence, certainly one far wider than *g*. Not for nothing is his book on the triarchic theory entitled *Beyond IQ*, so far beyond, indeed, that much of it is not relevant to the psychometric concept of intelligence, so clearly defined by its factor loadings.

Chapter nine

EEG and intelligence

Electroencephalography, the study of the electrical activity in the brain, would appear to be a good candidate for elucidating the nature of the neurophysiology of intelligence. This has become a subject of formidable technical complexity, as Jensen (1980) points out, and my discussion of the findings in this area is indebted to the work of two friends and former colleagues, Tony Gale and Paul Barrett, both specialists in the field.

As was the case with factor analysis, before I examine the findings, some definitions of some of the technical terms will be given, although I shall keep my treatment of this topic as free from jargon as possible.

Definitions

EEG

This is the actual print-out of the electrical activity of the brain, as it is recorded from electrodes on the scalp of the subject. Studies of the EEG are generally of two types: spontaneous EEG, where electrical activity is recorded over periods of time generally greater than about 1 minute, and evoked potential EEG, where activity is recorded over very short periods of time, usually from about 0.25 to 2 seconds. This activity is evoked from the subject by presentation of well-defined stimuli or by requiring some cognitive performance or motor action.

EP, AEP and ERP

EP is the abbreviation for evoked potential; AEP is the average evoked potential to a number of stimuli, around 50 to 100 being required for reliable measurement. ERP stands for an event-related potential and is synonymous with EP.

In spontaneous EEG there are various wave forms, classified in terms of frequencies.

Alpha, gamma, delta and theta rhythms

Alpha rhythm occurs when we are relaxed, with eyes closed; beta waves may be seen when we are awake with no movement; gamma is another waking rhythm; and delta occurs when we are asleep. Theta is held to reflect stress or emotion. There are other spontaneous EEGs, such as the saw-toothed wave in rapid eye movement (REM) sleep, which is associated with dreaming, but these are the main wave forms.

EEG research

Before examining the research relating EEG to intelligence I shall summarise, briefly, the significance and psychological meaning of the EEG, as far as there is agreement. Thompson (1975) argues that essentially EEG activity can be seen as a continuum from fast amplitude waves when we are attentive and aroused, to the slower alpha waves when we are relaxed, to waves of lower and lower frequency, as we sleep more and more deeply. That rapid eye movement sleep shows all the EEG signs of high arousal presents no problem since the REM state is conceived as a third biological state (Snyder, 1965), sleep and wakefulness being the others. The EEG, then, can be seen as an index of brain activity.

I hope that this brief introduction to EEG research will be sufficient for an understanding of the work relating it to intelligence. First, a few general points need to be made. An obvious question arises as to the purpose of this kind of work. What useful information could be obtained, even if some characteristic EEG patterns could be discerned while individuals were solving intelligence test items or otherwise thinking intelligently? Again would it be useful if it could be shown that highly intelligent people had different EEGs from those of less intelligence?

EEG traces indicate brain activity of a general kind. They can give information concerning the location of brain damage or abnormal activity, but are far less accurate than brain scanning, or intrusive studies of neuronal activity with micro electrodes. This means that EEG studies could never pin down the exact neural substrates of intelligence, even if any such exist. At the most they might indicate the general areas of the brain involved in intelligent problem solving. This is not of great interest since it is assumed, in any but the most mystical psychologies, that neurological activity underlies cognitive performance. In this sense, therefore, studies showing a relationship between brain activity and intelligence are banal. It is inconceivable that there should be no relationship. Only findings of very high precision convey useful information; EEG studies, at least with current technology, are unable to give these.

The second question concerning the use of EEG traces as an index or test of intelligence is obviously of greater interest. Some investigators would undoubtedly like to use such an EEG measure as an intelligence test on the grounds that it was more than just the standard IQ test because it was unaffected by social class and education. While this is true in the sense that there is nothing specifically learned in the environment that directly affects the EEG, unlike the standard intelligence test item, this is an argument fraught with difficulties and shows the danger of sliding, without noticing, from the psychological into the physiological world of discourse. It is a difficult argument for the following reasons. First, let us assume that the correlation between IQ test and EEG measure is very high. Without this it could not be used as a measure. We know that social class and education affect IQ scores: this is why we want to eliminate their effects. We know that brain activity and IQ are correlated highly. If they were not, EEG could never be a measure of IQ. If they are, then brain activity, as measured by the EEG, must also be affected by social class and education. If it were not, it would not correlate with IQ. It follows, therefore, that an objective measure of IQ which is both free from the contamination of social influences and highly correlated with IQ is impossible. This is true not only of EEG but also of all objective indices, including reaction time and its variability and inspection time. If the IQ test is the criterion, the influences that contaminate that measure must influence the underlying physiology. Thus the search for an objective measure of intelligence is a chimera. If the EEG measure remains unaffected by social factors its correlation with IQ will not be large and it will need external validation.

Nevertheless despite these difficulties, there is a considerable body of research into the relationship between IQ and the EEG. It is worth examining because, in principle at least, it will tell us something about the underlying physiology of intelligence, knowledge that is necessary for a complete (by definition) understanding of the topic.

The relation between EEG and intelligence test scores

Gale and Edwards (1983) have a summary of this work which forms a useful basis for our discussion. Ellingson (1956) carried out a broad survey of the relationships between EEG and a variety of psychological phenomena, in which he concluded that there was no relationship between EEG alpha and intelligence. Vogel and Broverman (1964) disagreed. They claimed that when studies of feeble-minded children, normal children, normal adults, the aged and those with brain disease were included, a different conclusion may be drawn, namely that there is a relationship between alpha and mental age. Studies with IQ, they claimed, were distorted by the failure to control for differences in chronological age. It should be noted, therefore, that two apparently authoritative reviews of essentially the same field arrive at the opposite conclusion. This argument was continued in Ellingson (1966) and Vogel and Broverman (1966), a dispute which is generally considered to have gone against Ellingson.

As Gale and Edwards (1983) point out this disagreement resulted from a number of procedural problems in the study of EEG and intelligence which are still relevant today and these should be mentioned before we come to examine more modern work. A global IQ score is a gross measure which may obscure genuine differences between ability and the EEG. They suggest that different sub-scales of the WAIS should be used and possibly measures of crystallised and fluid ability. EEG correlations with these scales may differ with differences in electrode placement. Ideally measures of the main ability factors should be included in all studies.

Sampling is a problem. Gale and Edwards (1983) argue that the full ability range should be used and warn that extreme groups may differ in so many ways that inference is impossible. Furthermore, special groups may yield special effects. Certainly it is difficult to draw conclusions from studies where a variety of samples is used. Similarly the effects of age must be controlled for or partialled out and separate results for males and females are required.

The other experimental demands are more general but are worth noting because the fact that they must be mentioned indicates the low quality of some of the research in this area. If subjects are performing tasks these should be relevant to intelligence tests. Experimenters should not know the IQs of subjects while recording or scoring the EEG. Elaborate analyses are required for determining the relationship between EEG parameters and IQ. These EEG parameters should be reliable and obtained from multiple channel recordings. Finally caution has to be exercised in the interpretation of results, especially with respect to attention. Since attention is necessarily involved in both IQ tests and the experimental tasks it might be argued that what has been found is an EEG correlate of attention. Experiments must be designed to minimise this inferential problem.

One point to emerge from this discussion of the experimental difficulties of studies of the EEG and intelligence is that the original investigations of spontaneous EEG can hardly meet these criteria and that ERP studies are superior. This is the reason, of course, that most of the recent work in this field falls into this second category; nevertheless some work with spontaneous EEG is valuable and this will be examined.

As Barrett and Eysenck (in press) argue, an excellent investigation was that of Bosaeus *et al.* (1977), who studied a sample of 138 normal children aged between 5 and 16 years using computerised broad spectral band EEG parameters derived from multichannel recordings. The WISC was the measure of intelligence and age was partialled out of all correlations: a study therefore which met most of the methodological requirements suggested by Gale and Edwards (1983). The highest correlation was –0.21 between parieto-occipital theta and WISC IQ. A series of studies by Gasser and colleagues, discussed by Barrett and Eysenck (in press), showed a correlation of about 0.5 between spectral power and WISC IQ among children, the reliability of the EEG measures being around 0.7 (Gasser *et al.*, 1985).

Giannitrapani (1985) has published a monograph in which the relationships between WISC and WAIS scores and spectral power at various electrode replacements were reported. Here it was shown that performance of the various subtests of the Wechsler scales is related to EEG power at specific frequencies at specific brain areas. It is generally accepted that the investigations conducted by Giannitrapani, over the years, were of a very high quality technically, and it establishes beyond doubt that spontaneous EEG is related to IQ, although in a complex way. However, as Gale and Edwards argue (1983) some caution over some of the absolute values is necessary since these are extremely high.

Evoked potential EEG and intelligence

I shall now examine the research into evoked potential EEG and intelligence, which, as the power of computers has increased, has become the main approach to this question. In the last twenty years a body of systematic evidence has accumulated demonstrating relationships between AEP parameters and IQ test scores. As was discussed at the beginning of this chapter, the AEP is usually based upon 50 to 100 evoked potentials, so large a number being necessary because the signal to noise ratio is poor. Although EPs can be taken from any position of electrodes for work with intelligence the Cz, central, position is usually employed.

Ertl and his colleagues (e.g. Chalke and Ertl, 1965; Ertl, 1971, 1973) were among the first to report correlations between AEP and IQ. As Barrett and Eysenck (in press) argue, the average correlation across these and other studies is around –0.3, although some investigators were unable to replicate these findings (e.g. Davis, 1971).

Ertl (1971, 1973) and Shucard and Callaway (1973) computed the power spectra of the AEPs and found that the amplitude of the maximum spectral frequency did not correlate significantly with IQ, although, as Barrett and Eysenck (in press) point out, some investigators did find significant correlations.

It is obvious from these two paragraphs that there were conflicting results in this field. Gale and Edwards (1983) discuss some of the technical problems in these experiments and it is evident that no clear, simple account of the results is possible. However, there is, fortunately, some more recent (and controversial) work in this field and it is not necessary to try to resolve these conflicting findings.

Hendrickson and Hendrickson (1980, 1982) and Blinkhorn and Hendrickson (1982) carried out a series of studies of the correlation between AEP and IQ. Two measures were extracted from the AEPs. A complexity measure was derived from measuring the contour perimeter of the AEP wave form – the string measure. The larger this value the higher the IQ because it was supposed to be an index of the detail in which the stimulus was encoded through the nervous system. The second measure, the variance, was negatively correlated with IQ. This was calculated from each sample point on the AEP, over a number of epochs. Variability was held to reflect poor fidelity of transmission.

Blinkhorn and Hendrickson (1982) using a sample of 33 polytechnic students used the string measure with auditory AEPs. This was correlated with scores on verbal tests and the Advanced Raven's Matrices. The verbal tests did not correlate significantly with the string score but the correlation with the Matrices was 0.45. However, this was boosted by correcting it for attenuation due to restriction of range on the matrices and this yielded a correlation of 0.84, which was close to what the Hendricksons (1980) obtained when they reanalysed some of the Ertl data using their string measure, although it should be noted that this study had used extreme scores which tend to distort correlations. Nevertheless correlations around 0.8 are remarkable.

In the second study (Hendrickson and Hendrickson, 1982), 219 children were given the Wechsler test; scale scores, verbal IQ, performance IQ and total IQ were correlated with the string and variance measures, and a composite EEG score, the difference between these two. The correlations with the total IQ were 0.72 (string), –0.72 (variance) and –0.83 (composite). Those with the performance and verbal scores were similar but a little lower. In sum these were extraordinary findings. Haier *et al.* (1983) found a substantial correlation of 0.45 for a string measure of a visual AEP and Raven's Matrices, supporting these findings. Before discussing this work another approach needs to be described.

These are studies of the effects of information processing on AEPs. Here, of course, if the cognitive task resembles the items of intelligence tests we might argue that any findings were the biological correlates of intelligence, far more so than with the AEPs to simple stimuli. There have been a number of such studies and those by Schafer (e.g. 1979, 1982; Schafer and Marcus, 1973) are typical. In general it is found that unexpected stimuli produce AEPs of a higher amplitude than do expected stimuli. Schafer has argued that high IQ subjects expend less effort in processing expected stimuli and thus show lower amplitude AEPs than less intelligent controls. This difference in effort, which he labels neural adaptability, correlates about 0.6 with IQ. This figure when corrected for attenuation due to homogeneity rises to 0.78. Schafer (1984, 1985) has modified this work in answer to the criticism that his neural adaptability might simply reflect habituation. Similar high correlations were found.

Finally mention must be made of the work of Robinson, who adds in formidable complexity by attempting to relate the findings to Pavlov's notions of excitatory and inhibitory processes. This work (Robinson, 1982) is too speculative for me to describe it here but interested readers should consult his original papers.

What are we to make of all these findings? The first point to note is the extraordinarily high correlations reported. These are so high that I suspect some artefactual error. Psychology only rarely yields correlations of this magnitude. The study by the Hendricksons has attracted considerable attention and discussion because, if the figures were taken at their face value, their string measure would be an almost perfect test of intelligence.

However, there are some serious problems with this work, some of which have been discussed by Barrett and Eysenck (in press). If the AEP were regarded as a measure of biological intelligence and if it were perfectly valid and reliable it follows that it could not correlate more highly with IQ tests than the square root of the heritability of the IQ. Now this is estimated, as we have seen, to be around 60 per cent and this means that the highest possible correlation would be 0.77. However, since no biological measure could be perfectly reliable or cover all the biological variance the highest correlation to be expected would be only about 0.65. This line of reasoning makes it highly unlikely that the string measure could be measuring biological intelligence.

Another problem is that the wave forms reported by the Hendricksons and by Ertl are unlike those usually found with AEPs and the suspicion that there is some artefact in this work is further attested by the fact that Barrett was unable to replicate, despite taking the most elaborate care, either the Hendrickson or the Schafer results.

However, what Barrett did find was a far more convincing and moderate correlation between intra-subject variability in AEP and IQ of between 0.4 and 0.5. Barrett and Eysenck (in press) regard this as a sound and replicable finding.

Thus this, in my view, is what has to be interpreted – the fact that there is a correlation of about 0.45 between IQ and variability in AEP. Barrett and Eysenck (in press) relate this work to the research that we reviewed in Chapter 7 on speed of mental processing and IQ. They argue that these findings represent the same phenomenon, namely that IQ is related to the error free transmission of information through the cortex, errors being responsible for low IQ, variability in reaction time and in AEP, and generally slow transmission of information through the system.

Conclusions

This is a highly technical field. Many differences affect results. Electrode placement; automated Fourier analysis of AEPs and other measurement methods; stimuli to evoke the EEGs, whether they are complex or simple; methods of presentation, auditory or visual, intense or weak, and the order in which they are given; the intelligence measures, global or analytic, fluid or crystallised, verbal or non-verbal; sampling extreme groups or the full range of ability. As was described, as yet there is no set of standard procedures in this work which allows comparability of results. Nevertheless there does seem to be a firm finding, in this work, that AEP correlates around 0.45 with IQ, when suspect studies are discounted.

The work on cognitive processing and EAPs by Schaffer is undoubtedly one approach to the investigation of the underlying psychophysiology of intelligence which should be followed up in the future. As it stands it is difficult to know what to make of the results, given the problems of replication.

In summary I think that it can be argued from all this work that the findings support a claim that speed of information transmission is related to intelligence, although the argument that these are identical is certainly not supported. Clearly, on the present evidence, the notion of using some kind of EEG as a measure of intelligence is ruled out and this ignores the conceptual difficulty, which was discussed earlier in this chapter, as to whether such an objective measure would ever be possible. This is an area of research where much remains to be discovered.

Chapter ten

The nature of *g*

In this final chapter I am now in a position to describe the nature of *g* as it has been revealed by the psychometric studies which I have discussed in this book. What is portrayed here is the psychometric view of intelligence. It must be pointed out that almost all aspects of this description are supported by clear, unequivocal evidence, all of which has been discussed in the relevant sections of this book. It is essential that all this evidence is considered together. Each part, on its own, may be, although usually with little plausibility, explained without reference to intelligence, in some *ad hoc* way. However, when all the arguments are considered together these *ad hoc* explanations appear incoherent and are, of course, in violation of the law of parsimony which in most sciences is a useful guide in selecting among competing hypotheses. Finally when this description has been completed, I shall be able to rebut the criticisms of the psychometric concept of *g*.

The *g* factors

The first plank in the argument are the *g* factors. As was shown in Chapter 3 when the factor analysis of tests of ability is conducted according to the best technical criteria, two correlated *g* factors emerge, crystallised and fluid ability. Other solutions, such as those of Guilford, which yield large numbers of factors, were shown to be artefacts of their factor analytic procedures. The designation of certain factor analytic methods as technically adequate was not arbitrary or because they yielded factors which fitted the psychometric case. It was because in examples where the structure was known on other grounds these methods were capable of revealing it.

Thus we can feel confident that two factors, fluid and crystallised ability, account for a considerable portion of the variance in tests of ability. Intelligence can be thought of as an amalgam of these two factors. The more controversial work of Undheim (1981) which identifies fluid ability with Spearman's *g* makes little difference to the argument, except that greater emphasis would be placed on fluid ability than on crystallised ability.

Description of the *g* factors

The next point in the argument turns on the nature of these *g* factors, as they are defined by their factor loadings. First we must stress their ubiquity. It is exceedingly difficult to construct a differential aptitude test, for example, because usually it turns out that the tests are so highly loaded on the *g* factors that they are not differential, merely poor tests of intelligence.

The factor loadings make the nature of these factors quite explicit. Both are concerned with reasoning, the ability to educe correlates, an ability which is valuable in the solution of problems in almost all spheres of human activity. The more a task or test item demands such a capacity the more highly it loads on these *g* factors.

The difference between these two factors is clear, although they are positively correlated and were not distinguished until the late 1930s after factor analysis had been developed by Thurstone. Fluid ability is the basic reasoning ability which is a function, in the main, of our neural structures. Crystallised ability is fluid ability, as it is evinced in the culture, in the abilities and skills which are valued in a culture. This differs, therefore, across place and time. In Victorian England, for example, fluid ability was invested in the classics. Today the computing sciences have replaced them. In a culture totally different from that of the West crystallised ability must also differ, hence the problems of cross-cultural testing.

Fluid ability, by definition, has to be tested, either by tasks with which subjects are entirely unfamiliar or which they all know completely, such as letter and number sequences. Testing both fluid and crystallised ability shows that the two are most highly correlated in infancy. As children grow up and become affected by their experiences at home and school, so the two abilities diverge. Of course fluid ability sets a limit to the intellectual capacity of an individual which the finest education cannot overcome. Very bright children who have been culturally disadvantaged often invest their abilities in skills which are not valued in the culture and may not appear successful at school or in occupations. In general, however, in a society where there is good educational opportunity, regardless of race, class or wealth, there will be a reasonably high correlation between these two *g* factors. Indeed where there is not, the disparity may be a good index of the inequality of a society. As was argued above, intelligence may be thought of as an amalgam of these two factors.

Other ability factors

A third important aspect of human abilities can be seen in the factor analyses which were discussed in Chapter 3. The *g* factors are the largest factors in terms of the variance for which they account. However, there are other factors and their structure can be seen at both the first and second order level, that is as rather narrow primary factors or as a few broader, more general secondary factors. Full lists were given in Chapter 3. Gardner's (1983) notion of multiple intelligences was shown to be misconceived in the light of the factor analytic results.

To understand the importance of these other factors the psychometric model of performance should be recalled where successsful performance is seen as a function of various abilities, not just intelligence, together with personality, motivational, and mood factors which are outside the scope of this book. Thus it is no counter to the psychometric emphasis on intelligence to argue that brilliant musicians or artists are not the highest scorers on intelligence tests. In fact there are a number of specific factors, such as musical ability, and group factors, such as spatial and verbal ability, which may be more important than intelligence for certain special tasks. However, what has been shown is that if we want to predict performance for any job, the nature of which was quite unknown to us, the test most likely to be successful would be a test of one or both of the *g* factors.

Intelligence tests

It was shown that there exist highly efficient tests of both crystallised and fluid abilities. The pre-factorial tests of intelligence, the WAIS, WISC and the Stanford-Binet, measure both *gf* and *gc*. It is better to measure these separately. For adults Raven's Matrices and the Culture-Fair Test are good tests of fluid ability. For crystallised ability Miller's

Analogies is effective, at least with educated people. The best single measure of crystallised ability is the vocabulary test.

The most common criticism of intelligence tests is that they measure only skill at intelligence tests, a view mistakenly thought to be supported by the definition of intelligence as what intelligence tests measure. In terms of the psychometric model of tests this claim means that the variance in intelligence tests is specific. This is entirely wrong. If it were so, intelligence tests would not correlate with other tests or with any criteria at all. In fact in the field of abilities intelligence tests correlate positively with almost all abilities and with a wide variety of real-life criteria. Thus, whatever it is that intelligence tests measure it is not peculiar to intelligence tests.

Indeed the nature of these correlations with external criteria, their pattern and size, is one of the best ways of defining intelligence, and these correlations will now be discussed. In appraising these results it should be remembered that most intelligence tests measure both the *g* factors.

Correlations with educational criteria

Although the absolute size of correlations can be misleading, due to restriction of range so that, for example, the correlation between IQ and university degree class is only about 0.2, whereas for the whole range of ability it would rise to around 0.6, there is no doubt that intelligence tests are good predictors of educational performance at all ages. Traditional, academic subjects load more highly on *g*, those that are found to be difficult, than do subjects such as dance studies and youth in society.

Opponents of IQ testing cannot deny these facts but attempt to explain them in terms of social class or that the IQ test is nothing but an attainment test or even that there is some further variable, perhaps one of motivation, that accounts for the performance on both IQ and educational tests. All these arguments have been shown to be empty. With social class partialled out there is still a substantial correlation between IQ and educational performance, as the original Northumberland Tests showed. The argument derived from the claim that IQ is an attainment test is destroyed by the fact that these tests predict performance in subjects in which the testees have had no previous experience. If there were some motivational variable that predicted educational attainment, it is a strange fact that no measures of motivation correlate well with educational attainment, and studies of attainment using multiple regression indicate clearly that the personality and motivational variance is quite separate from the ability variance. Thus these objections will not do. Crystallised and fluid ability are highly involved with academic attainment.

Correlations with occupational criteria

In more than 10,000 studies the average correlation of IQ with occupational success was 0.3 (so much for the claim that IQ is an attainment test or measures only ability at IQ tests). Given the problems of restriction of range and the difficulty of measuring occupational success, this correlation is certainly a low estimate of its true size. It must be pointed out, in addition, that no other variable, either of ability or personality, can approach this figure. It implies that being intelligent is advantageous to doing any job (no correlations were negative), considerably so in many cases, usually those jobs which are regarded as difficult, requiring clever people to carry them out. This, of course, is not to deny the importance of other ability factors in occupational success as well as personality factors. We are not arguing that intelligence is the sole determinant of job success.

Taken together, this and the previous section delineate intelligence, factorially defined and measured by intelligence tests, in a mould similar to the popular conception – as a variable which contributes to success at school and in a job and which discriminates clever people from the rest.

Heritability of intelligence

Biometric studies of intelligence test scores are unequivocal: approximately 65 per cent of the population variance in intelligence is attributable to genetic factors. However, there are some important points to be remembered when reflecting on this finding. The fact that Burt falsified his results does not mean that his claims must be false. They are supported by all the other research. The results apply to the variance within a population so that they should not be extrapolated to different populations. In those where the range of environments was very large, the genetic contribution could well be smaller.

These findings have interesting implications for our understanding of intelligence. First the fact that there is a large genetic component does not mean that environmental influences are negligible. Even with the highest genetic estimates environmentallly produced differences can be as large as 30 points. There is no doubt, therefore, that the environment is important in the development of intelligence. Biometric studies have shown that it is the non-shared family experiences that contribute to the variance in intelligence test scores. This presents a problem for intervention, as many such projects indicate. It is interesting to note that the best single predictor of an individual's score on an intelligence test is the mean of his or her parents' scores, regardless of whether he or she was brought up by them.

That there is a large genetic component in intelligence test scores again makes the claim that these scores reflect only skill at the test difficult to believe. Furthermore the findings indicate that although the environment plays a significant part in the development of intelligence there is a limit to what can be achieved. Those writers who argue that all can be explained by opportunity and experience are refuted by biometric research. These studies support an unpopular truth. People are not born equal and if individuals fail it is not necessarily because their family was negligent or they have made no effort. Finally the problems of cross-cultural testing were discussed, difficulties so severe that comparison of the intelligence of different cultural groups is of dubious psychological meaning.

Summary

At this point it is useful to summarise the conclusions that can be drawn from the first seven sections of this chapter. So far intelligence, as measured by many intelligence tests, has appeared as an amalgam of two factors, crystallised and fluid ability, although ideally both should be measured separately. Far from being specific to these tests these factors are implicated in almost all cognitive skills and correlate well with educational and occupational achievement. In addition there is a considerable genetic component to their variance such that important although the environment is in the development of intelligence, especially the non-shared family environment, ultimately there is a limit, set by our genes, to what we may achieve.

In the next sections of this chapter I shall outline the studies of what mental processes may be involved in intelligence and I shall discuss the research into its underlying neurophysiology.

Mental speed and intelligence

The relation of mental speed to intelligence has been subjected to intensive investigation in recent years, especially by Jensen, whose claims were shown to be exaggerated. Hick's law states that speed of responding to stimuli is a function of their complexity, in terms of bits of information. Unfortunately this is not a universal law since at least 20 per cent of subjects do not fit it. Even among those who do the Hick parameters, the slope and intercept of the regression line are correlated with intelligence only to a moderate extent, around 0.2. The slope, it will be recalled, is interpreted as the index of the time required for the mental processes of attention, sensory registration of the stimuli, trans- mission of the signal to the brain, encoding of the stimuli, transmission of the response signal and muscle lag. The intercept is the estimate of the time involved in discrimination and choice. Apart from the fact that not all subjects obey Hick's law, in some samples even these moderate correlations are not found. The more simple measures of mean reaction time, and especially variability of reaction time, do correlate moderately, about 0.3 with intelligence, but as the choices become more complex these correlations do not increase, as they would if mental speed were an important aspect of intelligence. All these findings make it difficult to support Jensen's conclusions that the work with choice reaction time and intelligence implicates the importance for intelligence of mental speed in a limited channel capacity system.

Similar claims have been made for the research on the correlation between inspection time and intelligence. In small samples with a huge range of IQs, from retardate to highly intelligent, some very high correlations were reported, in some cases higher than the reliability of the measures. However, psychometrically adequate research shows that the correlation of IT with IQ is very small, again around 0.3.

There can be no doubt: IT cannot be used as an objective measure of intelligence and the results cannot be used as a support for any strong theoretical claims linking intelligence and mental speed.

Intelligence and cognitive processes

Much modern work in the field of abilities has concerned itself with the study of the cognitive processes underlying the ability factors. It was hoped that this would explicate their psychological nature. Carroll (1980) carried out extensive studies of elementary cognitive tasks

(ECTs) which were held by cognitive psychologists to measure basic cognitive processes. However, examination of these tasks showed that in many cases they were little different from primary factors but of an extremely narrow kind.

Sternberg and his colleagues, in numerous publications (e.g. Sternberg, 1985a), have been the most zealous workers in this field and Sternberg has produced a triarchic theory of intelligence which extends the meaning of intelligence well beyond the psychometric concept. This work is best known for its componential analysis of abilities, these components being essentially cognitive processes. It was demonstrated that these components, despite the huge amount of research devoted to them, were non-contingent concepts, that is that they were a priori and entailed in the meanings of the problem-solving they were intended to explain. Their status was independent of empirical evidence. Their postulation, therefore, if it is a discovery of any note, is a tribute to philosophy rather than science. It is not clear that these components have greatly added to our knowledge of the *g* factors.

Finally we examined the work of Hunt and his colleagues, pioneers in this field. Their main finding is that intelligence correlates moderately with a cognitive process known as speed of lexical accesss, speed of getting meaning from words. It would appear that this gives people more time to carry out other mental processes which are required in problem solving.

It can be concluded that the attempt to tie cognitive psychology to psychometrics has not yielded much of great interest, which is the reason why Sternberg has so extended the notion of intelligence in his triarchic theory.

EEG and intelligence

Studies of spontaneous EEG and intelligence, as demonstrated by the technically sound work of Giannitrapani (1985), indicate that there is a relationship between intelligence and EEG, although it is complex and difficult to characterise in a simple way. It means little more than that brain activity, as measured by the EEG, differs in individuals of differing intelligence. It would be strange if this were not so.

Studies of evoked potentials or event-related potentials reveal one well-established point, namely that variability of response is negatively related to intelligence. The correlations are substantial, about 0.45, but fall far short of the somewhat unlikely claims made by some investigators, notably the Hendricksons. This level of correlation rules out absolutely the hope of using EEG as an objective measure of intelligence. This would be possible only if the correlations with IQ tests were in the order of 0.9 or if it could be shown that the EEG measures were more valid than the IQ tests. Neither of these possibilities seems to be near at hand.

However, this substantial and replicated correlation between the variability of EAPs and intelligence requires explanation. The best one seems to be that of Barrett and Eysenck (in press), who regard it as further evidence that high intelligence is associated with speed of information transmission through the nervous system, an interpretation similar to that given by Jensen (1987a) to the results of his reaction time studies, although such an interpretation has been challenged in this book.

Conclusions: the psychometric view of intelligence

In the previous sections of this final chapter I have presented the main findings and interpretations of the psychometric work on intelligence. I have done this so that, as I draw my conclusions, the facts on which they are based may be in front of us.

The factor analysis of tests of ability unequivocally yields two *g* factors labelled by Cattell (1971) as fluid and crystallised intelligence. Fluid intelligence appears to be our basic reasoning ability, dependent ultimately on the neural efficiency of our brains. Crystallised ability is the set of skills, valued by our culture, in which this ability is invested. It is idle to argue that these intelligence factors are in some way artefacts of the tests, measuring nothing but skill at the tests, because these factors correlate with a wide variety of external criteria as we shall discuss below. Indeed it is difficult to construct tests of ability that do not implicate these *g* factors.

A study of what variables load on these *g* factors and the relative size of the loadings allows them to be described with complete precision, unlike simple verbal definitions of intelligence. Thus factorial studies indicate that the *g* factors are implicated in tests and tasks where reasoning is required. The *g* factors fit closely the original Spearman definition of intelligence as the ability to educe correlates. The definition of intelligence as what intelligence tests measure is not as circular as it first appears because what intelligence tests measure is clearly defined by their factor loadings.

Of course the psychometric view of intelligence does not state that the intelligence factors are the only ability factors. There are a large number of well-established factors and there can be no definitive list. However, modern factor analytic research has managed to structure these factors into a meaningful hierarchy. The two intelligence factors are generally considered to account for the largest portion of the ability variance, being ubiquitous, for all abilities are positively correlated, but other smaller broad second order factors, such as retrieval and visualisation which are important in certain skills, have been reliably established. At the primary level there are more than twenty clear factors (but far less than those required by the Guilford model, which is infirmed by poor rotational methods) and these are important in certain special skills. Music ability is an obvious example. In this psychometric model of abilities, then, performance in any task is seen as a function of the relevant ability factors and among these the *g* factors are always important but to varying extents. Ability and motivational factors also occur in this model, it should be noted, but are irrelevant to this book. Finally it should be noted that some modern factor analytic research equates fluid ability with Spearman's *g*.

An important aspect of science is not only the postulation of concepts but also the ability to measure them. Vague concepts, immune to measurement, are not useful. Since the psychometric notion of intelligence was derived from the factor analysis of tests it is not surprising that there are excellent tests of intelligence. In terms of factor loadings (tests loading on the crystallised or fluid intelligence and no other factors), Raven's Matrices and their associated vocabulary scales are excellent, as are the Cattell intelligence tests. The WISC scales, which were developed before modern, refined factor analysis, are still useful measures although they load on both *g* factors. In brief, there is no doubt that psychometric intelligence, crystallised and fluid ability, can be measured with great reliability and validity. Indeed the measurement of intelligence is probably the highest technical achievement of psychology, up to this date.

The fact that intelligence can be measured with high reliability and validity means that intelligence tests have been extensively used in applied psychology. There can be no doubt that intelligence tests are excellent predictors of both occupational and academic achievement. Indeed Vernon (1961) argued that the prediction of academic success through the use of IQ tests in the 11+ selection system was about as efficient as could reasonably be expected of a test. The fact that on average intelligence correlated 0.3 with success in any job speaks for itself. This is powerful evidence in support of *g* as a basic reasoning factor.

Howe (1988) has attempted to explain away these results in two ways. On the one hand he argues that the IQ tests simply measure the achievement measured by the educational criterion. This is simply wrong since IQ tests are excellent at predicting success in subjects quite unknown to those taking the tests. His other explanation is equally contradicted by evidence when he claims that there must be some common factor, such as persistence, underlying both academic success and the IQ score. Correlations with personality variables show this argument to be false as do the factor analyses where no such higher order factor emerges.

That this reasoning ability measured by the *g* factors is a basic human characteristic, rather than some fortunate skill at an arbitrary set of items, is attested by the work on the heritability of IQ. Here it can be shown that in about 65 per cent of the population variance in intelligence is attributable to genetic factors. Furthermore, contrary to intuition, the important aspect of the environmental determinant is the non-shared family environment. These findings indicate that both genetic and environmental factors are important in the development of intelligence and to ignore either is absurd. However, it does mean that those concerned with education must accept that there are natural limits to individuals' ability.

Attempts to show more precisely, in terms of specified mental processes, just what the *g* factors are, have not met outstanding success. Despite the claims of Jensen (1987a) intelligence cannot be equated with speed of transmission of information through the nervous system. Nor does it appear to be highly dependent upon it, although the studies with AEPs, where the variability of AEPs is correlated with IQ, might be interpreted in this way. Nor can Sternberg's experimental work be said to have advanced much beyond the stage of careful thinking about the nature of problem solving since his components can be shown to be a priori processes.

Nevertheless intelligence is a useful and powerful construct. It can be measured with high reliability and validity and the measures can be used to predict performance in a large number of real-life settings although caution has to be shown in interpreting national differences in intelligence. Furthermore, the high heritability of these scores points to the fact that they reflect some basic underlying physiological substrata as do the correlations with the AEP. In addition to attribute, causal status to intelligence fits in well with everyday observation. To study the lives of intellectual giants forces one to the conclusion that part of their success was attributable to their being intelligent. Certainly hard work, a good education and good opportunities are important but many thousands of individuals have these good fortunes and do nothing. For example, Lord Russell recalled how as a neglected child at the age of 7 his brother had introduced him to Euclid. Immediately, he writes, he could see the proof and he demanded to know the evidence for the postulates. This is a perfect illustration of high *g* in action.

Such, then, is psychometric *g*. It is a concept that people used before experimental psychology was invented, which, given its ubiquity, is hardly surprising. What psychometrics has done is to define the concept more precisely and to provide good measures. In addition it has begun to sketch in its biological basis and to investigate its underlying mental processes. In the light of all the evidence, to write intelligence off as a redundant abstraction or as an instrument of middle-class oppression seems absurd.

References

Anastasi, A. (1961) *Psychological Testing*. New York, Macmillan.

Barrett, P. & Eysenck, H.J. (in press) *Brain Electrical Potentials and Intelligence*. London, Institute of Psychiatry.

Barrett, P., Eysenck, H.J. & Luching, S. (1989) Reaction time and intelligence: a replicated study. *Intelligence, 10*, 9–40.

Barrett, P. & Kline, P. (1981) The observation to variable ratio in factor analyses. *Journal of Personality and Group Psychology, 1*, 23–33.

Barrett, P. & Kline, P. (1982) Factor extraction: an examination of three methods. *Personality and Group Behaviour*, 2, 94–98.

Berry, J.W. & Dasen, P.R. (1974) *Culture and Cognition: Readings in Cross-Cultural Psychology*. London, Methuen.

Binet, A. & Simon, T. (1905) Méthodes nouvelles pour le diagnostic du niveau intellectual des anormaux. *L'Annee Psychologique, 11*, 191–244.

Blinkhorn, S.F. & Hendrickson, D.E. (1982) Average evoked responses and psychometric intelligence. *Nature, 295*, 596–597.

Boring, E.G. (1923) Intelligence as the tests test it. *New Republic, 35*, 35–37.

Bosaeus, E., Matusek, M. & Petersen, I. (1977) Correlations between paedo-psychiatric findings and EEG variables in well functioning children of ages 5–16 years. *Scandinavian Journal of Psychology, 18*, 140–147.

Brand, C.R. & Deary, I.J. (1982) Intelligence and inspection time. Chapter 5 in Eysenck, H.J. (Ed.) (1982).

Brown, F.G. (1976) *Principles of Educational and Psychological Testing* (2nd edn). New York, Holt, Rinehart & Winston.

Buros, O.K. (Ed.) (1978) *The VIII Mental Measurement Yearbook*. New Jersey, Gryphon Press.

Butcher, H.J. (1973) Intelligence and creativity. In Kline, P. (Ed.) (1973) *New Approaches in Psychological Measurement*. London, Wiley.

Carroll, J.B. (1980) Individual Difference Relations in Psychometric and Experimental Cognitive Tasks. Lab. Report. No. 163, University of North Carolina.

Carroll, J.B. (1983) Individual differences in cognitive abilities. Pp.213–235 in Irvine, S.H. & Berry, J.W. (Eds) (1983).

Cattell, R.B. (1966) The Scree test for the number of factors. *Multivariate Behaviour Research 1*, 140–161.

Cattell, R.B. (1967) The theory of fluid and crystallised intelligence. *British Journal of Educational Psychology, 37*, 209–224.

Cattell, R.B. (1971) *Abilities: Their Structure, Growth and Action*. New York, Houghton Mifflin.

Cattell, R.B. (1973) *Personality and Mood by Questionnaire*. San Francisco, Jossey Bass.

Cattell, R.B. (1978) *The Scientific Use of Factor Analysis in Behaviourial and Life Sciences*. New York, Plenum.

Cattell, R.B. (1981) *Personality and Learning Theory*. New York, Springer.

Cattell, R.B. & Butcher, H.J. (1968) *The Prediction of Achievement and Creativity*. New York, Bobbs-Merrill.

Cattell, R.B. & Cattell, A.K.S. (1959) *The Culture-Fair Test*. Champaign, Ill., IPAT.

Cattell, R.B. & Johnson, R.C. (Eds.) (1986) *Functional Psychological Testing*. New York, Brunner Mazel.

Cattell, R.B. & Kline, P. (1977) *The Scientific Analysis of Personality and Motivation*. London, Academic Press.

Chalke, F. & Ertl, J. (1965) Evoked potentials and intelligence. *Life Sciences, 4*, 1319–1322.

Child, D. (1970) *The Essentials of Factor Analysis*. London, Holt, Rinehart & Winston.

Chopin, B.H. (1976) Recent developments in item-banking, in De Gruitjer, P.N.M. & van der Kamp, L. (Eds) (1976).

Claridge, G. (1989) Special review. *Personality and Individual Differences, 10*, 1111–1112.

Cooper, C., Kline, P. & Maclaurin-Jones, L. (1986) Inspection time and primary abilities. *British Journal of Educational Psychology, 56*, 304–308.

Cronbach, L.J. (1984) *Essentials of Psychological Testing* (4th edn). New York, Harper & Row.

Cronbach, L.J. & Snow, R.E. (1977) *Aptitudes and Instructional Methods: A Handbook for Research on Interactions*. New York, Irvington.

Darlington, C.P. (1970) *Heredity, 25*, 655–656.

Davis, F.B. (1971) The measurement of mental ability through evoked potential recording. *Educational Record Research Bulletin I*.

De Gruitjer, P.N.M. & van der Kamp, L. (Eds) (1976) *Advances in Psychological and Educational Measurement*. London, Wiley.

Deregowski, J.B. (1980) *Illusions, Patterns and Pictures: A Cross-Cultural Perspective*. London, Academic Press.

Dixon, N.F. (1976) *On the Psychology of Military Incompetence*. London, McGraw-Hill.

Ekstrom, R.B., French, J.W. & Harman, H.H. (1976) *Manual for Kit of Factor-Referenced Cognitive Tests*. New Jersey, Educational Testing Service.

Ellingson, R.J. (1956) Brain waves and problems of psychology. *Psychological Bulletin, 53*, 1–34.

Ellingson, R.J. (1966) Relationships between EEG and test intelligence: a commentary. *Psychological Bulletin, 65*, 91–98.

Elliot, C. (1983) *The British Ability Scales*. Windsor, N.F.E.R.

Ericsson, K.A. (1988) Analysis of memory performance in terms of memory skill. Chapter 5 in Sternberg, R.J. (Ed.) (1988).

Erlenmeyer-Kimmling, L. & Janek, L.F. (1963) Genetics and intelligence. A review. *Science, 142*, 1477–1479.

Ertl, J. (1971) Fourier analysis of evoked potentials and human intelligence. *Nature, 230*, 525–526.

Ertl, J. (1973) Evoked potentials and Fourier analysis. *Nature, 241*, 209–210.

Eysenck, H.J. (1967) Intelligence assessment. A theoretical and experimental approach. *British Journal of Educational Psychology, 37*, 81–98.

Eysenck, H.J. (1980) *The Causes and Effects of Smoking*. London, Temple Smith.

Eysenck, H.J. (Ed.) (1982) *A Model for Intelligence*. New York, Springer.

Eysenck, H.J. & Eysenck, S.B.G. (1975) *The EPQ Test*. London, Hodder & Stoughton.

Eysenck, M.W. (1984) *A Handbook of Cognitive Psychology*. London, Erlbaum.

Feldman, M. & Lewontin, R. (1975) The heritability hang up. *Science, 190*, 1163–1168.

Fletcher, R. (1987) The doubtful case of Sir Cyril Burt. *Social Policy and Administration, 21*, 40–57.

French, J.W. (1951) The description of aptitude and achievement factors on terms of rotated factors. *Psychometric Monograph, 5*.

Fulker, D.W. (1979) Nature and nurture: heredity. Chapter 5 in Eysenck, H.J. *The Structure and Measurement of Intelligence*. New York, Springer Verlag.

Gale, A. & Edwards, J.A. (Eds) (1983) *Physiological Correlates of Human Behaviour* Vol. III. Chapter 6. Correlates of Intelligence. London, Academic Press, 74–97.

Galton, F. (1883) *Inquiries into Human Faculty and its Development*. London, Macmillan.

Garber, H.L. (1988) *The Milwaukee Project: Preventing Mental Retardation in Children at Risk*. Washington, DC, American Association of Mental Retardation.

Gardner, H. (1983) *Frames of Mind: The Theory of Multiple Intelligence*. New York, Basic Books.

Gasser, T., Bacher, P. & Steinberg, H. (1985) The test–retest reliability of spectral parameters of the EEG. *Electroencephalograph and Clinical Neurology, 60*, 312–319.

Ghiselli, E.E. (1966) *The Validity of Occupational Aptitude Tests*. New York, Wiley.

Giannitrapani, D. (1985) *The Electrophysiology of Intellectual Functions*. London, Karger.

Gottfredson, L.S. (1986) Societal consequences of the *g* factor in employment. *Journal of Vocational Behaviour, 29*, 379–410.

Guilford, J.P. (1958) *Psychometric Methods*. New York, McGraw-Hill.

Guilford, J.P. (1959) *Personality*. New York, McGraw-Hill.

Guilford, J.P. (1964) Zero correlations among tests of intellectual abilities. *Psychological Bulletin, 61*, 401–404.

Guildford, J.P. (1967) *The Nature of Human Intelligence*. New York, McGraw-Hill.

Guilford, J.P. & Hoepfner, R. (1971) *The Analysis of Intelligence*. New York, McGraw-Hill.

Gustaffson, J.E. (1988) Hierarchical models of individual differences in cognitive abilities. Chapter 2 in Sternberg, R.J. (Ed.) (1988).

Haier, R.J., Robinson, D.L. Braden, W. & Williams, D. (1983) Electrical potentials of the cerebral cortex and psychometric intelligence. *Personality and Individual Differences, 5*, 293–301.

Hakstian, A.R. (1971) A comparative evaluation of several prominent methods of oblique factor transformation. *Psychometrika, 36*, 175–193.

Hakstian, A.R. & Cattell, R.B. (1974) The checking of primary ability structure on a broader basis of performance. *British Journal of Educational Psychology, 44*, 140–154.

Hakstian, A.R. & Cattell, R.B. (1976) *Manual for the Comprehensive Ability Battery*. Champaign, Ill., IPAT.

Harman, H.H. (1976) *Modern Factor Analysis*. Chicago, University of Chicago Press.

Hernshaw, I.S. (1979) *Cyril Burt Psychologist*. London, Hodder & Stoughton.

Heim, A.W. (1975) *Psychological Testing*. London, Oxford University Press.

Heim, A.W., Watts, K.P. & Simmonds, V. (1970) *AH4, AH5 and AH6 Tests*. Windsor, N.F.E.R.

Hendrickson, D.E. & Hendrickson, A.E. (1980) The biological basis of individ- ual differences in intelligence. *Personality and Individual Differences, 1*, 3–33.

Hendrickson, D.E. & Hendrickson, A.E. (1982) The biological basis of intelligence, in H.J. Eysenck (Ed.) (1982).

Hick, W. (1952) On the rate of gain of information. *Quarterly Journal of Experimental Psychology, 4*, 11–26.

Horn, J. & Cattell, R.B. (1966) Refinement and test of the theory of fluid and crystallised intelligence. *Journal of Educational Psychology, 57*, 253–270.

Horn, J. & Knapp, J.R. (1973) On the subjective character of the empirical

base of Guilford's structure of intellect model. *Psychological Bulletin, 80,* 33–43.
Howe, M.J.A. (1988) Intelligence as an explanation. *British Journal of Psychology, 79,* 349–360.
Hudson, L. (1966) *Contrary Imaginations.* London, Methuen.
Humphreys, L.G. (1975) Addendum. *American Psychologist, 30,* 95–96.
Hunt, E.B. (1976) Varieties of cognitive power. In Resnik, R.B. (Ed.) (1976).
Hunt E.B. (1978) Mechanics of verbal ability. *Psychological Review, 85,* 109–130.
Irvine, S.H. (1969) Factor analysis of African abilities and attainments. *Psychological Bulletin, 71,* 20–32.
Irvine, S.H. & Berry, J.W. (Eds) (1983) *Human Assessment and Cultural Factors.* New York, Plenum.
Jenks, C. (1972) *Inequality: A Reassessment of the Effects of Family and Schooling in America.* New York, Basic Books.
Jensen, A.R. (1974) How biased are culture loaded tests? *Genetic Psychology Monographs, 90,* 185–244.
Jensen, A.R. (1980) *Bias in Mental Testing.* New York, Free Press.
Jensen, A.R. (1982) Reaction time and psychometric *g*. Chapter 4 in Eysenck, H.J. (Ed.) (1982).
Jensen, A.R. (1985) The nature of the black–white differences on various psychometric tests. Spearman's hypothesis. *Behaviourial and Brain Sciences, 8,* 193–219.
Jensen, A.R. (1987a) Individual differences on the Hick paradigm. In Vernon, P.A. (Ed.) (1987) *Speed of Information Processing and Intelligence.* Norwood, Ablen.
Jensen, A.R. (1987b) The *g* beyond factor analysis. In Royce, R.R., Glover, J.A. & Witt, J.C. (Eds.) (1987) *The Influence of Cognitive Psychology on Testing.* Hillsdale, NJ, Erlbaum.
Jensen, A.R. (1989) Raising IQ without increasing *g*? *Developmental Psychology.*
Jensen, A.R. *et al.* (1980) Précis of bias in mental testing. *Behaviourial and Brain Sciences, 3,* 325–371.
Jinks, J.L. & Fulker, D.W. (1970) Comparison of the biometrical, genetical, MAVA and classical approaches to the study of human behaviour. *Psychological Bulletin, 73,* 311–349.
Joreskog, K.G. (1969) A general approach to confirmatory maximum likelihood factor analysis. *Psychometrika, 34,* 183–202.
Kamin, L.J. (1974) *The Science and Politics of IQ.* Harmondsworth, Penguin.
Kline, P. (1975) *The Psychology of Vocational Guidance.* London, Batsford.
Kline, P. (1979) *Psychometrics and Psychology.* London, Academic Press.
Kline, P. (1986) *A Handbook of Test Construction.* London, Methuen.
Kline, P. (1987) Comments on Ronald Fletcher's defence of Cyril Burt. *Social Policy and Administration, 21,* 105–108.

Kline, P. (1988) *Psychology Exposed: The Emperor's New Clothes*. London, Routledge.

Kline, P. & Barrett, P. (1983) The factors in personality questionnaires among normal subjects. *Advances in Behaviour Research and Therapy, 5*, 141–202.

Kline, P., May, J. & Cooper, C. (1986) Correlations among elementary cognitive tasks. *British Journal of Educational Psychology, 56*, 111–118.

Krebs, E.G. (1969) The Wechsler Preschool and Primary Scale of Intelligence and Prediction of Reading Achievement in First Grade. PhD. Rutgers University (quoted by Jensen, 1980).

Lohnes, P.R. & Gray, M.M. (1972) Intelligence and the cooperative reading studies. *Reading Research Quarterly, 8*, 52–61.

Longstreth, L.E. (1984) Jensen's reaction-time investigations of intelligence: a critique. *Intelligence, 8*, 139–160.

Lord, F.M. (1974) *Individualised Testing and Item Characteristic Curve Theory*. Princeton, NJ, Educational Testing Service.

Lynn, R. (1987) The intelligence of the mongoloids: a psychometric, evolutionary and neurological theory. *Personality and Individual Differences, 8*, 813–844.

Lynn, R., Pagliari, C. & Chann, J. (1988) Intelligence in Hong Kong measured for Spearman's *g* and the neurospatial and verbal primaries. *Intelligence, 12*, 423–433.

McGeoch, J.A. (1942) *The Psychology of Human Learning*. New York, Longmans Green.

Mackintosh, N.J. (1981) A new measure of intelligence. *Nature, 209*, 529–530.

May, J., Kline, P. & Cooper, C. (1987) The construction and validation of a battery of tests to measure flexibility of thinking in army officers. APRE Working Papers WP15/87.

Miles, T.K. (1957) Contributions to intelligence testing and the theory of intelligence. 1. On defining intelligence. *British Journal of Educational Psychology, 27*, 153–165.

Miller, W.S. (1970) *Miller Analogies Test*. New York, Psychological Corporation.

Muller-Hill, B. (1988) *Murderous Science. Elimination by Scientific Selection of Jews, Gypsies and Others. Germany 1933–1945*. Oxford, Oxford University Press.

Nettelbeck, T. (1982) Inspection time: an index for intelligence. *Quarterly Journal of Experimental Psychology, 24A*, 299–312.

Nettelbeck, T. & Lally, M. (1976) Inspection time and measured intelligence. *British Journal of Psychology, 67*, 17–22.

Nunnally, J.O. (1978) *Psychometric Theory*. New York, McGraw-Hill.

Ortar, C.R. (1963) Is a verbal test cross-cultural? *Scripta hierosolymitana, 13*, 219–235.

Otis, A.S. (1954) *Quick Scoring Ability Tests*. London, Harrap.
Pedley, R.R. (1953) *The Comprehensive School*. Harmondsworth, Penguin.
Pellegrino, J.W. (1985) Inductive reasoning ability. Chapter 9 in Sternberg, R.J. (Ed.) (1985b).
Plomin, R. (1986) *Development Genetics and Psychology*. Hillsdale, NJ, Erlbaum.
Plomin, R. (1988) The nature and nurture of cognitive abilities. Chapter 1 in Sternberg, R.J. (Ed.) (1988).
Plomin, R. & Daniels, D. (1987) Why are children in the same family so different from each other? *Behaviourial and Brain Sciences, 10*, 1–16.
Posner, M.I. (1978) *Chronometric Explorations of Mind*. Hillsdale, NJ, Erlbaum.
Quereshi, M.Y. (1972) *The Differential Aptitude Test* in Buros, O.K. (Ed.) (1972) *VIIth Mental Measurement Yearbook*. New Jersey, Gryphon Press.
Rasch, G. (1960) *Probabilistic Models for some Intelligence and Attainment Tests*. Copenhagen, Denmark Institute of Education.
Raven, J.C. (1965a) *The Crichton Vocabulary Scale*. London, H.K. Lewis.
Raven, J.C. (1965b) *The Mill-Hill Vocabulary Scale*. London, H.K. Lewis.
Raven, J.C. (1965c) *Progressive Matrices*. London, H.K. Lewis.
Resnik, R.B. (Ed.) (1976) *The Nature of Intelligence*. Hillsdale, NJ, Erlbaum.
Robinson, D.L. (1982) Properties of the diffuse thalamocortical system, human intelligence, and differentiated vs integrated modes of learning. *Personality and Individual Differences, 3*, 339–405.
Roth, E. (1964) Die Geschwindigkeit der Verarbeitung von Information und ihr Zussammenhang mit Intelligenz. Z. Exp. Angew. Psychol. *11*, 616–622. Cited by Eysenck, H.J. (1967).
Royce, J.R. (1963) Factors as theoretical constructs. Chapter 24 in Jackson, D.N. and Messick, S. (Eds.) (1967) *Problems in Human Assessment*. New York, McGraw-Hill.
Russell, B. (1911) *Principia Mathematica*. Cambridge, Cambridge University Press.
Sarason, S.B., Davison, K.S., Lighthall, F.J., Waite, R.R. & Ruebush, B.K. (1960) *Anxiety in Elementary School Children*. New York, Wiley.
Scarr, S. & Carter-Saltzman, L. (1982) Genetics and intelligence. Chapter 13 in Sternberg, R.J. (Ed.) (1982).
Scarr, S. & Weinberg, R.A. (1978) The influence of 'family background' on intellectual attainment. *American Sociological Review, 43*, 674–692.
Schafer, E. (1979) Cognitive neural adaptability: a biological basis for individual differences in intelligence. *Psychophysiology, 16*, 199.
Schafer, E. (1982) Neural adaptability: a biological determinant of behavioural intelligence. *International Journal of Neuroscience, 17*, 183–191.
Schafer, E. (1984) Habituation of evoked cortical potentials: correlates with intelligence. *Psychophysiology, 5*, 597.

Schafer, E. (1985) Neural adaptability: a biological determinant of g factor intelligence. *Behaviourial and Brain Sciences, 8*, 2, 264–270.

Schafer, E. & Marcus, M.M. (1973) Self stimulation alters human sensory brain responses. *Science, 181*, 175–177.

Seashore, C.E. (1919) *The Psychology of Musical Talent*. New York, Burdett.

Shucard, D. & Callaway, E. (1973) Relationship between human intelligence and frequency of analysis of cortical evoked responses. *Perceptual and Motor Skills, 36*, 147–151.

Shuter, R. (1968) *The Psychology of Musical Ability*. London, Methuen.

Skinner, B.F. (1953) *The Science of Human Behaviour*. New York, Macmillan.

Smedslund, J. (1978) Some psychological themes are not empirical: reply to Bandura. *Scandinavian Journal of Psychology, 19*, 101–102.

Snow, R.E. & Yalow, E. (1982) Education and intelligence. Chapter 9 in Sternberg, R.J. (Ed.) (1982).

Snyder, F. (1965) Progress in the new biology of dreaming. *American Journal of Psychiatry, 122*, 377–392.

Spearman, S. (1904) 'General intelligence': objectively determined and measured. *American Journal of Psychology, 15*, 201–292.

Sternberg, R.J. (1977) *Intelligence, Information Processing and Analogical Reasoning: The Componential Analysis of Human Abilities*. Hillsdale, NJ, Erlbaum.

Sternberg, R.J. (Ed.) (1982) *Handbook of Human Intelligence*. Cambridge, Cambridge University Press.

Sternberg, R.J. (1984) Towards a theory of intelligence. *Behaviourial and Brain Sciences, 7*, 269–287.

Sternberg, R.J. (1985a) *Beyond IQ: A Theory of Human Intelligence*. Cambridge, Cambridge University Press.

Sternberg, R.J. (Ed.) (1985b) *Human Abilities: An Information-Processing Approach*. London, W.H. Freeman.

Sternberg, R.J. (1986) Inside intelligence. *American Scientist, 74*, 137–143.

Sternberg, R.J. (Ed.) (1988) *Advances in the Psychology of Human Intelligence* Vol. IV. Hillsdale, NJ, Erlbaum.

Sternberg, R.J. & Gardner, M.K. (1983) Unities in inductive reasoning. *Journal of Experimental Psychology: General, 112*, 80–116.

Sternberg, R.J. & Powell, J.S. (1963) Comprehending verbal comprehension. *American Psychologist, 38*, 878–893.

Sternberg, S. (1960) High speed scanning in human memory. *Science, 153*, 652–654.

Sternberg, S. (1975) Memory scanning: new findings and current controversies. *Quarterly Journal of Experimental Psychology, 27*, 1–32.

Terman, L.M. & Merrill, M.A. (1960) *Stanford-Binet Intelligence Scale*. New York, Houghton Mifflin.

Terman, L.M. & Oden, M. (1959) *The Gifted Group at Mid Life*. Stanford, California University Press.

Thompson, R.F. (1975) *Introduction to Physiological Psychology*. New York, Harper & Row.

Thomson, G.H. (1946) *Moray House Intelligence Test*. London, University of London Press.

Thorndike, R.L. & Hagen, E. (1959) *Ten Thousand Careers*. New York, Wiley.

Thurstone, L.L. (1947) *Multiple Factor Analysis: A Development and Expansion of Vectors of the Mind*. Chicago, University of Chicago Press.

Undheim, J.O. (1981) On intelligence II. A neo-Spearman model to replace Cattell's theory of fluid and intelligence. *Scandinavian Journal of Psychology*, *22*, 181–187.

Undheim, J.O. & Horn, J.L (1977) Critical evaluation of Guilford's structure of intellect theory. *Intelligence*, *1*, 65–81.

Velicer, W.F. (1976) Determining the number of components from the matrix of partial correlations. *Psychometrika*, *41*, 321–327.

Vernon, P.A. (Ed.) (1987) *Speed of Information Processing and Intelligence*. Norwood, Ablen.

Vernon, P.E. (1960) *Intelligence and Attainment Tests*. London, University of London Press.

Vernon, P.E. (1961) *The Measurement of Abilities*. London, University of London Press.

Vernon, P.E. (1969) *Intelligence and Cultural Environment*. London, Methuen.

Vernon, P.E. (1979) *Intelligence, Heredity and Environment*. New York, W.H. Freeman.

Vogel, W. & Broverman, D.M. (1964) Relationships between EEG and test intelligence: a critical review. *Psychological Bulletin*, *62*, 132–144.

Vogel, W. & Broverman, D.M. (1966) A reply to 'Relationship between EEG and test intelligence: a commentary'. *Psychological Bulletin*, *65*, *99–109*.

Wahlsten, P. (1980) Race, the heritability of IQ and the intellectual scale of nature. In Jensen *et al*. (1980) 358–359.

Waller, J.A. (1971) Achievement and social mobility. Relationships among IQ score, education and occupation in two generations. *Social Biology*, *18*, 252–259.

Warburton, F.W. (1951) The intelligence of the Gurkha recruit. *British Journal of Psychology*, *42*, 123–133.

Warburton, F.W. (1965) Observations on a sample of psychopathic American criminals. *Behaviour Research Therapy*, *3*, 129–135.

Wechsler, D. (1944) *Measurement of Adult Intelligence* (3rd edn). Baltimore, Williams & Wilkins.

Wechsler, D. (1958) *The Measurement and Appraisal of Adult Intelligence* (4th edn). Baltimore, Williams & Wilkins.

Wechsler, D. (1974) *Manual for the Wechsler Intelligence Scale for Children* revised. New York, Psychological Corporation.
Wechsler, D. (1975) Intelligence defined and undefined: a relativistic appraisal. *American Psychologist, 30*, 135–139.
Wing, H.D. (1936) *Tests of Musical Ability in School Children.* M.A. Thesis, University of London.
Wissler, C. (1901) The correlation of mental and physical trials. *Psychological Monographs, 3*, 1–62.
Witkin, H.A. (1962) *Psychological Differentiation.* New York, Wiley.

Index